NANOTECHNOLOGY SCIENCE AND TECHNOLOGY

APPLIED ASPECTS OF NANO-PHYSICS AND NANO-ENGINEERING

NANOTECHNOLOGY SCIENCE AND TECHNOLOGY

Additional books and e-books in this series can be found on Nova's website under the Series tab.

NANOTECHNOLOGY SCIENCE AND TECHNOLOGY

APPLIED ASPECTS OF NANO-PHYSICS AND NANO-ENGINEERING

KIRILL LEVINE
AND
ANDREY SYRKOV
EDITORS

nova
science publishers
New York

Library of Congress Cataloging-in-Publication Data

ISBN: 978-1-53614-708-7

Published by Nova Science Publishers, Inc. † New York

CONTENTS

PREFACE

This volume of the series "Nanotechnology Science and Technology" entitled "Applied Aspects of Nanophysics and Nano-engineering" is partially composed of Proceedings of International Symposium "Nanophysics and Nano-Engineering 2017" (venue: Mining University) (Section one), and extended manuscripts, related to the same area of expertise.

A variety of phenomena were described. Smart nanostructured coatings, methods of synthesis based on both "top to bottom" (plasma deposition, remote methods) and "bottom to top" approaches were covered, as well as modeling approaches and analytical techniques. As before, ecological issues were highly addressed, such as materials for water purification and pollution prevention.

Permanent interest to fullerenes as to one-dimensional carbon-based structure arises from their ability to be relatively easily modified by species of interest, for the purpose of bio-substrate delivery (Semenyuk, Keskinov). Graphite exfoliation was utilized as a method to produce graphite nanoparticles (Raksha). Modelling of fullers was reported by Barbin.

Issues of dielectric relaxation of solids have been a stunning topic as a research method continuously, for at least few decades, and even now the interest to dielectric relaxation approach seems to even increase. This is because of sensitivity of this non-destructive method to conformational changes of flexible molecular moieties, brushes, interchain segments. This avenue was actively expanded by Nikonorova (focused on materials appliances of method) and Castro (technical development of method and resolution, as well as materials study).

Semiconductor technologies discussed in the book were related to developing solar concentrator systems (silicon technologies) (Moshnikov), heterojunction solar cells (Maslov) eutectic gallium arsenide solid solutions for development of alternative heterostructures based on tunneling effect (Ivanova). "Exotic" semiconductors –

diamonds with delta-doped layers known for their high temperature resistance were studied by capacitance measurements (Shestakova).

Directional crystallization was studied to produce rear-Earth compounds with anisotropic properties for application in thermoelectric materials (Nemov).

Findings in sorption properties of clay minerals was reported by Pak. Role of singlet oxygen is underestimated as global in environmental factor. Paper by Nazarenko reports studied in oil shale and oil shale ash Baltic basin.

Materials with magnetic properties synthesized by sol-gel method were based on varrium-titanium ceramic were studied by variety of powerful experimental methods: SEM, XRD, SAXC, SAPNS (Shilova).

Tomaev and Pshchelko reported findings in surface modification of zinc oxide films modified by selenium. Special experimental setup made possible using ambient pressure approach without isolating from atmosphere to synthesize hierarchically ordered surface structure.

Interface properties related to water absorption on aluminum surface were reported by Syrkov, as they are of interest for tribology applications of organopolymer compositions.

Composite nanostructured materials for solar concentrator systems were discussed by Moshnikov and Mustafaev, compounds for thermionic energy converters containing cesium intercalated with graphite were reported by Mustafaev.

It is believed that this book provides an unbiased sketch of progress of science in nanotechnology and related areas, and would give pleasant moments of reading to specialist and interested reader.

In: Applied Aspects of Nano-Physics …
Editors: Kirill Levine and Andrey Syrkov
ISBN: 978-1-53614-708-7

Chapter 1

SYNTHESIS, IDENTIFICATION AND PHYSICAL-CHEMICAL PROPERTIES OF BIS-ADDUCT OF FULLERENE C60 WITH HISTIDINE

Vlada V. Petrenko[1], Nikolay A. Charykov[1,2], Konstantin N. Semenov[1,3], Viktor A. Keskinov[1,*], Aleksey V. Kurilenko[1], Irina V. Semenyuk[1] and Nikolay M. Safiannikov[2]

[1]Saint-Petersburg State Technological Institute (Technical University), Russia
[2]Saint-Petersburg State Electro-Technical University (LETI), Russia
[3]Saint-Petersburg State University, Russia

ABSTRACT

The heterogeneous – non-catalytic method of the synthesis and purification (by water – methanol re-crystallization) of bis-adduct of fullerene C_{60} with amino-acid histidine is described. Identification of bis-adduct was provided by the following methods of physical-chemical analysis: C-H-N-O Element analysis, High performance liquid chromatography, IR-spectroscopy, Electronic spectroscopy. Volume properties of bis-adduct of fullerene C_{60} with histidine water solutions - $C_{60}(C_6N_3H_8O_2)_2(H)_2$ – H_2O at 298 K (concentration dependencies of density, average and partial molar volumes of both components) were determined with the help of quartz pycnometers. Concentration dependencies of refraction properties (refraction indexes, specific and molar refractions

* Corresponding Author's Email: keskinov@mail.ru.

of the solutions and bis-adducts) at 298 K were determined with the help of refractometers. Using method of isothermal saturation in ampules, poly-thermal solubility of $C_{60}(C_6N_3H_8O_2)_2(H)_2$ in water solutions was determined in the temperature range 293 – 353 K. By the method of visible light scattering, concentration dependencies of linear dimensions of nano-cluster associates and electro-kinetic ζ – potentials were determined at 298 K. By the method of cryo-metry with the help of Beckmann thermometer concentration dependence of the decrease of liquid temperatures was determined at the temperatures nearby 273 K. On the base of last data excess function of both components - $C_{60}(C_6N_3H_8O_2)_2(H)_2$ andH_2O (activities and activity coefficients) were calculated, using novel semi-empirical Virial Decomposition Asymmetric Model (VD-AS). Also with the help of VD-AS diffusional instability concentration region was calculated.

Keywords: bis-adduct, fullerene C_{60}, histidine, synthesis, heterogeneous – non-catalytic method, identification, physical-chemical analysis, water solutions, physical-chemical properties.

INTRODUCTION

Water soluble derivatives of fullerenes is a perspective class of compounds due to possibilities of their application in various fields of science and technology especially in biology and medicine due to well definedmembranotrophic, cytoprotective, radioprotective, antioxidant, antimicrobial, antiviral and transport properties [1–3]. This chapter is devoted to the development of the cycle of works, devoted to the synthesis, identification and the investigation pf physical-chemical properties of the adducts of light fullerenes (C_{60} or C_{70}) and amino-acids: alanine, lysine, histidine, arginine, glycine, proline, etc. Stromet prepared fullerene amino acids by dipolar addition to C_{60} of either the Boc- or Fmoc-Na-protected azido amino acids derived from phenylalanine and lysine [4]. It was determined that prepared amino acids are a mixture of 5,6-open (major product) and 6,6-closed (minor product) derivatives. Kotelnikova et al. synthesized the water-soluble C60 derivatives with DL-alanine and investigated the influence of the obtained compounds on structure and permeability of the lipid bilayer of phosphatidylcholine liposomes [5]. Hu et al. synthesized the C60 derivative with β-alanine, cystine and arginine [6]. Kumar et al. synthesized and studied DNA cleavage efficiency upon visible light (in the presence of nicotinamide adenine dinucleotide) of the C60-lysine derivative [7]. Jiang et al. synthesized novel water-soluble C60-glycine derivative [8]. The results of cytotoxicity assay of cancer cell lines showed thatC60-glycine derivative in a dose-dependent manner increases cell death. Authors of [9, 10] synthesized hybrid structures based on fullereneC60 with attached proline amino acid (methyl ether of N-[(β-alanylhistidyl-ethyl) fullerenyl] proline, methyl ether of N-[(nitroxyethyl)fullerenyl] proline, methyl ether of N-[(2′,3′-dinitroxy-propyl)fullerenyl] proline, methyl ether of N-[mono-hydro-fullerenyl] proline, carnosine). It was determined that all studied compounds possess anantioxidant activity and inhibited glutamate induced Ca^{2+} uptake into synap to somes of the rat brain cortex. Some articles were fulfilled with the participation of the authors of the presented chapter [11-17].

MATERIALS AND METHODS

The heterogeneous – non-catalytic method of the synthesis of bis-adduct of fullerene C_{60} with amino-acid histidine -$C_{60}(C_6H_8N_3O_2)_2(H_2)$ or 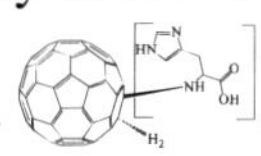 is described. Histidine (mass m = 3.2 g), base NaOH (m = 15.8 g) were dissolved in 54 cm^3 of H_2O; 130 cm^3 of the solution of fullerene C_{60} in o-xylene with the concentration 7.62 g/$dm^3$was prepared by direct dissolution. Then ethanol (C_2H_5OH) (m=267 g) was added to the both "o-xylene" and "water" phases and the heterogeneous mixture was stirred for 7 days. Then phases were divided, "water" phase was evaporated and bis-adduct $C_{60}(C_6H_8N_3O_2)_2(H_2)$ was precipitated by methanol (CH_3OH). Purification of bis-adduct was provided by triple recrystallization from water solutions by methanol. Additionally bis-adduct $C_{60}(C_6H_8N_3O_2)_2(H_2)$ was washed from the impurities of sodium-contain forms by HCl acid in Soxhlet apparatus. Yield of bis-adduct was 76% from the theoretical one.

Identification

Identification of bis-adduct $C_{60}(C_6H_8N_3O_2)_2(H_2)$ was provided by the following methods of physical-chemical analysis:

- C-H-N-O Element analysis (CHNS/O Analyzer 2400 Series II) proved formula of crystal hydrate $C_{60}(C_6N_3H_8O_2)_2(H)_2$*24$H_2O$;
- High performance liquid chromatography (HPLC PerkinElmer) confirmed purity of adduct 96.5 mass %;
- IR-spectroscopy (IRTracer-100, Shimadzu) in wavenumbers: ν = 400 – 4000 cm^{-1} demonstrates absorption characteristic peaks for: valent O-H, C-H, CH_2, C=O, deformation O-H, N-H oscillations and long-wave characteristic oscillations of C_{60} fullerene core (5009 – 1000 cm^{-1});
- Electronic spectroscopy (UV-1800, Shimadzu), wavelength: λ = 190-1100 nm demonstrates no effects and monotonically increasing with wavelength decreasing light absorption.

EXPERIMENTAL SECTION

Volume properties of bis-adduct of fullerene C_{60} with histidine water solutions - $C_{60}(C_6N_3H_8O_2)_2(H)_2 - H_2O$ at 298 K (concentration dependencies of density, average and partial molar volumes of both components) were determined with the help of quartz pycnometers. Concentration dependencies of refraction properties (refraction indexes, specific and molar refractions of the solutions and bis-adducts) at 298 K were determined with the help of refractometers. Using method of isothermal saturation in ampules, poly-thermal solubility of $C_{60}(C_6N_3H_8O_2)_2(H)_2$ in water solutions was determined in the temperature range 293 – 353 K. By the method of visible light scattering, concentration dependencies of linear dimensions of nano-cluster associates and electro-kinetic ζ – potentials were determined at 298 K. By the method of cryo-metry with the help of Beckmann thermometer concentration dependence of the decrease of liquid temperatures was determined at the temperatures nearby 273 K. On the base of last data excess function of both components - $C_{60}(C_6N_3H_8O_2)_2(H)_2$ andH_2O (activities and activity coefficients) were calculated, using novel semi-empirical Virial Decomposition Asymmetric Model (VD-AS). Also with the help of VD-AS diffusional instability concentration region was calculated.

ACKNOWLEDGMENTS

This work was supported by Russian Found of Fundamental Investigations – RFFI (Projects no. 16-08-01206, 18-08-00143 A), by the Grant of the President of Russian Federation for supporting young scientists MK- 4657.2015.3. Research was performed with using the equipment of the Resource Centers 'GeoModel,' Center for Chemical Analysis and Materials Research and Center for Thermogravimetric and Calorimetric Research of Research park of Saint Petersburg State University.

REFERENCES

[1] Cataldo, F., T. Da Ros, *Carbon Materials: Chemistry and Physics: Medical Chemistry and Pharmacological Potential of Fullerenes and Carbon Nanotube*, Springer, 2008.

[2] Sidorov, L. N., M. A. Yurovskaya, *Fullerenes, Ekzamen*, Moscow, 2005.

[3] Piotrovskii, L. B., O. I. *Kiselev, Fullerenes in Biology*, 2006 (Rostok, Saint-Petersburg).

[4] Strom, T. A., A. R. Barron, *Chem. Commun.* 46 (2010) 4764–4766.

[5] Kotelnikova, R. A., A. I. Kotelnikov, G. N. Bogdanov, V. S. Romanova, E. F. Kuleshova, Z. N. Parnes, M. E. Volpin, *FEBS Lett.* 389 (1996) 111–114.

[6] Hu, Z., W. Guan, W. Wang, L. Huang, X. Tang, H. Xu, Z. Zhu, X. Xie, H. Xing, *Carbon* 46. (2008) 99–109.

[7] Kumar, A., M. V. Rao, S. K. Menon, *Tetrahedron Lett.* 50 (2009) 6526–6530.

[8] Jiang, G., F. Yin, J. Duan, G. Li, *J. Mater. Sci. Mater. Med.* 26 (2015) 1–7.

[9] Grigoriev, V. V., L. N. Petrova, T. A. Ivanova, R. A. Kotelnikova, G. N. Bogdanov, D. A., Poletayeva, I. I. Faingold, D. V. Mishchenko, V. S. Romanova, A. I. Kotel'nikov, S. O. Bachurin, *Biol. Bull.* 38 (2011) 125–131.

[10] Tatyanenko, L. V., O. V. Dobrokhotova, R. A. Kotelnikova, D. A. Poletayeva, D. V., Mishchenko, I. Y. Pikhteleva, G. N. Bogdanov, V. S. Romanova, A. I. Kotelnikov, *Pharm. Chem. J.* 45 (2011) 329–332.

[11] Matuzenko, M. Y., D. P. Tyurin, O. S. Manyakina, K. N. Semenov, N. A. Charykov, K. V. Ivanova, V. A. Keskinov, *Nanosystems: Phys. Chem. Math.* 6 (2015) 704–714.

[12] Matuzenko, M. Y., A. A. Shestopalova, K. N. Semenov, N. A. Charykov, V. A. Keskinov, *Nanosystems: Phys. Chem. Math.* 6 (2015) 715–725.

[13] Shestopalova, A. A., K. N. Semenov, N. A. Charykov, V. N. Postnov, N. M. Ivanova, V. V. Sharoyko, V. A. Keskinov, D. G. Letenko, V. A. Nikitin, V. V. Klepikov, I. V. Murin, *J. Mol. Liq.* 211 (2015) 301–307.

[14] Manyakina, O. S., K. N. Semenov, N. A. Charykov, N. M. Ivanova, V. A. Keskinov, V. V., Sharoyko, D. G. Letenko, V. A. Nikitin, V. V. Klepikov, I. V. Murin, *J. Mol. Liq.* 211. (2015) 487–493.

[15] Semenov, K. N., N. A. Charykov, I. V. Murin, Y. V. Pukharenko, *J. Mol. Liq.* 201 (2015). 50–58.

[16] Semenov, K. N., N. A. Charykov, V. N. Keskinov, *J. Chem. Eng. Data* 56 (2011) 230–239.

[17] Semenov, K. N., N. A. Charykov, I. V. Murin, Y. V. Pukharenko, *J. Mol. Liq.* 201 (2015). 1–8.

In: Applied Aspects of Nano-Physics …
Editors: Kirill Levine and Andrey Syrkov
ISBN: 978-1-53614-708-7

Chapter 2

TRIS-ADDUCT OF LIGHT FULLERENE C_{70} WITH INDISPENSABLE AMINO-ACID LYSINE

Irina V. Semenyuk[1], Konstantin N. Semenov[1,3], Nikolay A. Charykov[1,2], Viktor A. Keskinov[1], Aleksey V. Kurilenko[1], Vlada V. Petrenko[1] and Nikolay M. Safiannikov[2]

[1]Saint-Petersburg State Technological Institute (Technical University), Saint-Petersburg, Russia
[2]Saint-Petersburg State Electro-Technical University (LETI), Saint-Petersburg, Russia
[3]Saint-Petersburg State University, Saint-Petersburg, Russia

ABSTRACT

The heterogeneous – non-catalytic method of the synthesis and purification (by water – methanol re-crystallization) of tris-adduct of fullerene C_{70} with amino-acid lysine is described. Identification of tris-adduct was provided by the following methods of physical-chemical analysis: C-H-N-O Element analysis, High performance liquid chromatography, IR-spectroscopy, Electronic spectroscopy, Complex thermal analysis. Volume properties of tris-adduct of fullerene C_{70} with lysine water solutions - $C_{70}(C_6H_{13}N_2O_2)_3(H_3)$at 298 K (concentration dependencies of density, average and partial molar volumes of both components) were determined with the help of quartz pycnometers. Concentration dependencies of refraction properties (refraction indexes, specific and molar refractions of the solutions and bis-adducts) at 298 K were determined with the help of refractometers. By the method of visible light scattering, concentration dependencies of linear dimensions of nano-cluster associates and electro-kinetic ζ – potentials were determined at 298 K. By the method of cryo-metry with the help of Beckmann thermometer concentration dependence of the decrease of liquid temperatures was determined at the temperatures nearby 273 K. On the base of last data excess function of both components - $C_{70}(C_6H_{13}N_2O_2)_3(H_3)andH_2O$ (activities and activity coefficients) were calculated, using novel semi-empirical Virial Decomposition Asymmetric Model (VD-AS). Concentration dependencies of density, dynamic and kinematic viscosity at 293, 313 and 333 K was determined by the Stokes method of "falling gold ball" in water solutions of tris-adduct at 298 K.

Keywords: tris-adduct, fullerene C_{70}, lysine, synthesis, heterogeneous – non-catalytic method, identification, physical-chemical analysis, water solutions, physical-chemical properties

INTRODUCTION

It is well-known, that light fullerenes (C_{60} and C_{70}) are practically are insoluble in water or water solutions or fluids, such as: physiological solutions, blood, lymph, CSF, gastric juice, food and wine products etc. This fact considerably limited application of light fullerenes and their derivatives in biology, medicine, pharmacology, cosmetics, food industry etc. despite of the fact that the last ones possess a number of very useful and rare properties, in particular: photo-protective, cytoprotective, radioprotective, antioxidant, bactericide, antiviral and transport. A wide range of articles were devoted to the synthesis, identification, physical-chemical, biological and medicine investigation in multicomponent systems with the adducts of light fullerenes with interchangeable and indispensable amino-acids (see, for example articles [1-14]).

MATERIALS AND METHODS

The heterogeneous – non-catalytic method of the synthesis of tris-adduct of fullerene C_{70} with amino-acid lysine - $C_{70}(C_6H_{13}N_2O_2)_3(H)_3$is described. Histidine (mass m = 4.5g), base NaOH (m = 15.8g) were dissolved in 54 cm^3 of H_2O; 130 cm^3 of the solution of fullerene C_{70} in o-xylene with the concentration 8.9 $g/dm^3$was prepared by direct dissolution. Then ethanol (C_2H_5OH) (m = 250g) was added to the both "o-xylene" and "water" phases and the heterogeneous mixture was stirred for 7 days. Then phases were divided, "water" phase was evaporated and tris-adduct $C_{70}(C_6H_{13}N_2O_2)_3(H)_3$was precipitated by methanol (CH_3OH). Purification of tris-adduct was provided by triple recrystallization from water solutions by methanol. Additionally tris-adduct $C_{70}(C_6H_{13}N_2O_2)_3(H)_3$was washed from the impurities of sodium-contain forms by HCl acid in Soxhlet apparatus. Yield of tris-adduct was 79% from the theoretical one.

EXPERIMENTAL SECTION

Identification

Identification of tris-adduct $C_{70}(C_6H_{13}N_2O_2)_3(H)_3$ was provided by the following methods of physical-chemical analysis:

- C-H-N-O Element analysis (CHNS/O Analyzer 2400 Series II) proved formula of crystal hydrate $C_{70}(C_6H_{13}N_2O_2)_3(H)_3*18H_2O$;
- High performance liquid chromatography (HPLC PerkinElmer) confirmed purity of tris-adduct 98.6 mass %;
- IR-spectroscopy (IRTracer-100, Shimadzu) in wavenumbers: $\nu = 400 - 4000\ cm^{-1}$ demonstrates absorption characteristic peaks for: valent O-H, C-H, CH_2, C = O, deformation O-H, N-H oscillations and long-wave characteristic oscillations of C_{60} fullerene core ($5009 - 1000 cm^{-1}$);
- Electronic spectroscopy (UV-1800, Shimadzu), wavelength: λ = 190-1100 nm demonstrates no effects and monotonically increasing with wavelength decreasing light absorption;
- Complex thermal analysis shows stepwise external dehydration of crystal hydrates at 90 - 210°C, stepwise decarboxylation, internal dehydration, denitrogenation at 300 - 810°C, C_{70} fullerene core oxidation at the temperatures higher than 900°C.

Physical-Chemical Properties

Volume properties of tris-adduct of fullerene C_{70} with lysine water solutions - $C_{70}(C_6H_{13}N_2O_2)_3(H_3)$ at 298 K (concentration dependencies of density, average and partial molar volumes of both components) were determined with the help of quartz pycnometers. Concentration dependencies of refraction properties (refraction indexes, specific and molar refractions of the solutions and bis-adducts) at 298 K were determined with the help of refractometers. By the method of visible light scattering, concentration dependencies of linear dimensions of nano-cluster associates and electro-kinetic ζ – potentials were determined at 298 K. By the method of cryo-metry with the help of Beckmann thermometer concentration dependence of the decrease of liquid temperatures was determined at the temperatures nearby 273 K. On the base of last data excess function of both components - $C_{70}(C_6H_{13}N_2O_2)_3(H_3)$ andH_2O (activities and activity coefficients) were calculated, using novel semi-empirical Virial Decomposition Asymmetric Model (VD-AS). Also with the help of VD-AS diffusional instability concentration region was calculated. Concentration dependencies of density, dynamic and kinematic viscosity at 293, 313 and 333 K was determined by the Stokes method of "falling gold ball" in water solutions of bis-adduct at 298 K. Concentration dependencies of hydrogen indicator (with the help of glass electrode), specific electro-conductivity (by RLC meter) were determined in tris-adduct water solutions at 298 K. On the base of both last data seeming dissociation degrees and concentration dissociation constants were calculated. Potential-metric titration of - $C_{70}(C_6H_{13}N_2O_2)_3(H_3)$ water solutions by acid (HCl) and base (NaOH) were provided.

ACKNOWLEDGMENT

This work was supported by Russian Found of Fundamental Investigations – RFFI (Projects no. 16-08-01206, 18-08-00143 A), by the Grant of the President of Russian Federation for supporting young scientists MK- 4657.2015.3. Research was performed with using the equipment of the Resource Centers 'GeoModel', Center for Chemical Analysis and Materials Research and Center for Thermogravimetric and Calorimetric Research of Research park of Saint Petersburg State University.

REFERENCES

[1] Gan, Liang Bing, Luo, Chu Ping. *Chinese Chemical Letters,* 1994. Vol. 5, No. 4, p. 275 - 278.

[2] Sidorov, L. N., Yurovskaya, M. A. *Fullerenes*, Ekzamen, Moscow, 2005.

[3] Kumar, A., Rao, M. V., Menon, S. K. *Tetrahedron Lett.,* 50 (2009) 6526 - 6530.

[4] Kotelnikova, R. A., Kotelnikov, A. I., Bogdanov, G. N., Romanova, V. S., Kuleshova, E. F., Parnes, Z. N., Volpin, M. E. *FEBS Lett.,* 389 (1996) 111 - 114.

[5] Hu, Z., Guan, W., Wang, W., Huang, L., Tang, X., Xu, H., Zhu, Z., Xie, X., Xing, H. *Carbon,* 46. (2008) 99 - 109.

[6] Piotrovskii, L. B., Kiselev, O. I. *Fullerenes in Biology*, 2006 (Rostok, Saint-Petersburg).

[7] Jiang, G., Yin, F., Duan, J., Li, G. *J. Mater. Sci. Mater. Med.,* 26 (2015) 1 - 7.

[8] Manyakina, O. S., Semenov, K. N., Charykov, N. A., Ivanova, N. M., Keskinov, V. A., Sharoyko, V. V., Letenko, D. G., Nikitin, V. A., Klepikov, V. V., Murin, I. V. *J. Mol. Liq.,* 211. (2015) 487 - 493.

[9] Semenov, K. N., Charykov, N. A., Murin, I. V., Pukharenko, Y. V. *J. Mol. Liq.,* 201 (2015). 1 - 8.

[10] Semenov, K. N., Charykov, N. A., Keskinov, V. N. *J. Chem. Eng. Data,* 56 (2011) 230 - 239.

[11] Shestopalova, A. A., Semenov, K. N., Charykov, N. A., Postnov, V. N., Ivanova, N. M., Sharoyko, V. V., Keskinov, V. A., Letenko, D. G., Nikitin, V. A., Klepikov, V. V., Murin, I. V. *J. Mol. Liq.,* 211 (2015) 301 - 307.

[12] Semenov, K. N., Charykov, N. A., Murin, I. V., Pukharenko, Y. V. *J. Mol. Liq.,* 201. (2015). 50 - 58.

[13] Semenov, K. N., Iurev, G. O., Ivanova, N. M., Postnov, V. N., Sharoyko, V. V., Prikhodko, I. V., Murin, I. V., Charykov, N. A., Keskinov, V. A., Letenko, D. G., Kulenova, N. A. *J. Mol. Liq.,* 225. (2017). 767 - 777.

[14] Matuzenko, M. Yu., Shestopalova, A. A., Semenov, K. N., Charykov, N. A., Keskinov, V. A. *Nanosys.: Ph., Ch., Math.,* 6. (2015). № 5. 715 - 725.

In: Applied Aspects of Nano-Physics …
Editors: Kirill Levine and Andrey Syrkov
ISBN: 978-1-53614-708-7

Chapter 3

DIELECTRIC RELAXATION AND MOLECULAR MOBILITY IN THERMOPLASTIC AROMATIC POLYIMIDES WITH DIFFERENT STRUCTURES

***Natalia A. Nikonorova*[1], *Rene A. Castro*[2], *Dao T. Hong*[2,*] *and Alexei A. Kononov*[2]**
[1]Institute of Macromolecular Compounds
of Russian Academy of Sciences, Saint Petersburg, Russia
[2]Herzen State Pedagogical University of Russia, Saint Petersburg, Russia

ABSTRACT

The results of the study of the dielectric relaxation and molecular mobility in five thermoplastic aromatic polyimides with different structures are reported. Three regions of dielectric loss due to relaxation of dipole polarization: the γ and β (in the glassy state) and the α (in the highly elastic state) were found. The key dielectric relaxation parameters are determined for different temperature ranges.

Keywords: thermoplastic aromatic polyimides, dielectric relaxation, molecular mobility

INTRODUCTION

Thermoplastic aromatic polyimides (PIs) are high performance engineering polymers that are used in a wide range of specialty applications and have been studied extensively

---

* Corresponding Author's Email: honghip2012@gmail.com.

using dynamic mechanical and dielectric analysis. The wide possibilities for changing the chemical structure of the diamine and dianhydride components of the thermoplastic aromatic PIs allow one to change their physical properties. In this work, the dielectric spectroscopy method is used to study the dielectric relaxation and molecular mobility in five thermoplastic aromatic PIs of different structures.

Experimental Section

Dielectric spectra were measured in the frequency range from 10^{-1} Hz to 10^{6} Hz and the temperature range from -100°C to +400°C (in increments of 10°C and accuracy of 0.5°C) using "Concept 81" spectrometer (Novocontrol Technologies GmbH & Co). The experimental data included impedance (imaginary and real parts) measurements for the cell with the sample [1].

Results and Discussion

For all samples, the dielectric spectra, $\varepsilon^{//} = \varphi(f)$, show three regions of $\varepsilon^{//}_{max}$ due to relaxation of dipole polarization: the γ and β (in the glassy state) and the α (in the highly elastic state). At temperatures exceeding the α process, we observed a region of $\varepsilon^{//}_{max}$ corresponding to a conductivity relaxation.

The analysis of dielectric spectra was performed by describing complex permittivity ε^{*} with the empirical Havriliak-Negami (HN) equation (1).

Relaxation time values for all processes at different temperatures calculated according to the HN equation testify that the $-\log\tau_{max} = \varphi(1/T)$ dependences for the γ and β process are linear and can be described by the Arrhenius equation which is typical of the local forms of molecular mobility. For the α process, dependences of $-\log\tau_{max} = \varphi(1/T)$ are nonlinear, which is characteristic of cooperative types of molecular mobility and its were reliably described by the empirical Vogel–Tamman–Fulcher (VTF) equation (2).

$$\varepsilon^*(\omega) - \varepsilon_{\infty} = \sum_{k=1}^{n} \mathrm{Im}\left[\frac{\Delta\varepsilon_k}{\{1+(i\omega\tau_{HN_k})^{\alpha_k}\}^{\beta_k}}\right] \tag{1}$$

$$\tau_{max} = \tau_0 \exp\left(\frac{B}{T-T_0}\right) \tag{2}$$

Comparing the data obtained earlier for linear thermotropic polyesters and PI of different structures, it can be assumed that the appearance of the β process reflects the limited rotations of the phenyl rings and the adjacent polar –O– groups. Parameters of the Arrhenius equation, $\log\tau_0$ и E_a, are ~20 s and ~32 kcal/mole, respectively, exceed those for local processes, which can be explained by some influence of intermolecular interactions. For the γ process, $\log\tau_0$ и E_a, are ~15 s ~11 kcal/mole, respectively, are typical for the local mobility mode. Concerning the molecular mechanism of the γ process, there is no unequivocal explanation. For the γ and β processes, the molecular mobility practically does not depend on the chemical structure.

For the α process, influence of chemical structures on the molecular mobility was estimated by value of glass transition temperature T_g. For linear polymers, T_g is mainly determined by intermolecular interactions between macromolecules. In PI, the main contribution to intermolecular interactions is made by the dispersion forces between the flat phenylene rings. The chemical structure of the investigated PI is selected in such a way that it is possible to determine the contribution of each element entering into the diamine and dianhydride component on T_g of different systems. So, the introduction of oxygen "bridges" –O– between phenylene rings into the diamine part makes macromolecule more flexible, the intermolecular interactions between macromolecules decrease, and as a result, decrease in T_g by 30^0C. T_g is also influenced by dipole-dipole interactions. It is shown that the replacement of the polar group –O– by $-SO_2-$ in the dianhydride part of the macromolecule gives a decrease in T_g by 40^0C. This means that a consistent change in the chemical structure of the thermoplastic aromatic PIs, allows us to establish a correlation between the important characteristics as T_g and dielectric relaxation.

Reference

[1] Nikonorova, N. A.; Didenko, A. L.; Kudryavtsev, V. V.; Castro, R. A.: *Journal of Non-Crystalline Solids*, Vol. 447, (2016), p. 117.

In: Applied Aspects of Nano-Physics …
Editors: Kirill Levine and Andrey Syrkov
ISBN: 978-1-53614-708-7

Chapter 4

Analysis of Micro- and Nano-Pores of Oil Shale and Oil Shale Ash

M. Yu. Nazarenko*, N. K. Kondrasheva and S. N. Saltykova
Department of Processing Mineral Raw Materials,
Mining University, Saint-Petersburg, Russia

Abstract

In this study, we investigated such properties of oil shale and oil shale ash as the total porosity, the value of the square of the specific surface distribution of micro-and nano-pores. The change in the total porosity of the oil shale occurs in four stages. In the first stage (25–200 °C), the porosity increases due to the liberation of water, carbon dioxide, and hydrogen sulfide. In the second stage (200–400 °C), the porosity declines because of the bituminization of the shale. In the third stage (400–600 °C) and the fourth stage (600–900 °C), the porosity increases on account of semi coking and the decomposition of carbonate minerals. The average diameter of the pores of oil shale is in the limit of 8,962 nm - 21,244 nm.

Keywords: oil shale, oil shale ash, pores, porousness, surface area, micropore, sorbents

Introduction

Oil shale contain kerogen—that is, transformed organic matter from higher plants and simple organisms. They differ from coal in that they form a large quantity of tar on semi coking (20–70%, in terms of the organic component) [1]. After processing oil

* Corresponding Author's Email: max.nazarenko@mail.ru.

shales, a large quantity of oil shale ash remains (up to 50 wt % of the initial shale); its storage consumes enormous areas [2]. Oil shale also provides less energy mass for mass than other fossil fuels, including petroleum and natural gas. Nevertheless, interest in processing oil shale has risen sharply in the last decade. Researchers assert that effective industrial use of oil shales demands the use of not only the organic component but also the mineral component, so as to reduce the waste generation. [1-2]. The composition and surface reactivity of oil shales and spent shales were studied in [3-4]. On that basis, it seems that oil shales and oil shale ash may be used as sorbents, since they resemble materials such as sand and zeolite in terms of their composition (content of calcium, silicon, and aluminum oxides).

One of the important properties of the sorbents is the porosity and the specific surface area. On this basis, the aim of this work is to study the porosity of oil shale and oil shale ash.

EXPERIMENTAL SECTION

We study oil shales from the Baltic Basin in the Leningradsk field (< 0,125 mm fraction) and also oil shale ash (< 0,125 mm fraction).

The change in oil shale total porosity on pyrolysis in a nitrogen atmosphere and in air (at 25–900 °C) is studied pycnometrically. A 2g sample is placed in the pycnometer, into which ethyl alcohol is poured. The sample is held at 25 °C for 120 min.

The specific surface area and micro- and nano- pores of oil shale and oil shale ash were conducted on the analyzer, specific surface NOVA3200e. The cell device was weighed on an analytical balance. The mass was recorded. The cuvette was placed in samples weighing 2 g. in cuvettes with samples were prepared by a vacuum and was heated to 1000C and aging at this temperature is 2 hours, thus, the pores of the samples were freed from possible moisture. After that, the cuvettes with samples were weighed, the weight recorded. The cuvette was again placed in the device, they created a vacuum and the cell was dipped in liquid nitrogen. When establishing temperature equilibrium between the liquid nitrogen and the cuvette with the sample cuvette was starting to dispensed gas is nitrogen, and with the help of pressure sensors recorded the absorption of nitrogen by the sample. Determination of the specific surface area was by the method of BET (Bruner-Emmett-teller) based on the fact that the solid surface is considered as a set of adsorption sites, and in a state of dynamic equilibrium speed condensing at the free space of the molecules of the gas phase equals the rate of evaporation of molecules from the occupied places.

CONCLUSION

Definitely that the total porosity of shale is 24%, oil shale ash is 38%. The change in the total porosity of the oil shale occurs in four stages:

- In the first stage (25–200 °C), the porosity increases on account of the liberation of water, carbon dioxide, and hydrogen sulfide.
- In the second stage (200–400 °C), the porosity declines, on account of the bituminization of the shale.
- In the third stage (400–600 °C) and the fourth stage (600–900 °C), the porosity increases on account of semi coking and the decomposition of carbonate minerals.

Porosity of oil shale after heat treatment in an air atmosphere than in a nitrogen atmosphere, due to the oxidizing atmosphere and the consequent formation of a larger amount of gaseous products. The increase of porosity at temperatures above 800-900 °C is due to the formation of a large number of silicates. For example, in the conversion of alpha-quartz into alpha-cristobalite increases by 15.7%, which provides the change in porosity.

It is established that oil shale ash (51,796 sm^2/g) has a specific surface area higher than that of oil shale (9,427 sm^2/g), which is mainly due to the fact that during heat processing, a large amount of volatile components thereby increasing the porosity.

The average diameter of the pores of oil shale is in the limit of 8,962 nm - 21,244 nm.

REFERENCES

Bityukova L., Motler R. Composition of oil shale ashes from pulverized firing and circulating fluidized-bed boiler in Narva thermal power plants. *Oil Shale*. 2010. vol. 27. № 4. 339–353.

Leimbi Merike R., Tiina H., Eneli L., Rein K. Composition and properties of oil shale ash concrete. *Oil Shale*. 2014. Vol. 31, №. 2. 147–160.

Nazarenko M.Yu., Bazhin V.Yu., Saltykova S.N., Konovalov G.V. Physicochemical properties of fuel shale. *Coke and Chemistry*. 2014. №3. 44-49;

Nazarenko M.Yu., Kondrasheva N.K., Saltykova S.N. Surface reactivity of fuel shales from the Baltic basin. *Coke and Chemistry*. 2016. №5. 33-37.

Swift T., Mayer S. Study of thermal conversion of oil shale under N_2 and CO_2 atmospheres. *Oil Shale*. 2010. vol. 27. № 4. 309–320.

In: Applied Aspects of Nano-Physics …
Editors: Kirill Levine and Andrey Syrkov
ISBN: 978-1-53614-708-7

Chapter 5

STRUCTURE AND SORPTION PROPERTIES OF CLAY MINERALS

V. N. Pak*, Yu. Yu. Gavronskaya and E. S. Ivanova
Herzen State Pedagogical University of Russia,
Saint-Petersburg, Russia

ABSTRACT

A comprehensive study of the composition, structure, and sorption properties of a typical representative of clay minerals has been performed. The efficiency of extraction of Ni^{2+} ions from aqueous solutions with a concentration of 10^{-5} to 10^{-2} M has been determined. Such sorbent characteristics as chemical and mineralogical composition, specific surface, porosity, maximum sorption capacity, and number of active sites have been found to be mutually consistent.

Keywords: clay minerals, phase composition, surface, active centers, porous structure, ion

INTRODUCTION

Ion-exchange/sorption properties of natural clay minerals are determined by a combination of interrelated factors, such as chemical and mineralogical composition, crystallinity, number of active sites, porosity parameters, and conditions and results of chemical and thermal treatment. However, studies on sorption properties of particular

* Corresponding Author's Email: pakviacheslav@mail.ru.

samples often do not consider the complete set of the aforesaid factors, especially in adsorption–structural studies, substantiation of the solid/liquid phase ratio, and independent estimation of the number of surface active sites by different methods. As a result, a comparison of published data and prediction of sorption properties for particular samples become difficult [1-3]. The above stated also applies to some extent to solid surface studies. Nevertheless, fairly comprehensive and consistent description and analysis of interrelations of the above properties may be possible within the scope of conventional research practice.

MATERIALS AND METHODS

Typical clay samples of Russia north-west region were investigated. X-ray fluorescence analysis was performed with Shimadzu-13PK Instrument. Water adsorption was used to determine specific surface area and pore volume of clays. Sorption capacity of the samples towards Ni2+ was investigated as the decrease of ions content in solutions by the probes analysis with the use of EDTA.

RESULTS AND DISCUSSION

According to the X-ray fluorescence analysis data, the main components of the typical clay sample of Russia north-west region are the following oxides (wt %): SiO_2 (55.16), Al_2O_3 (15.52), CaO (6.51), Fe_2O_3 (5.33), MgO (2.73), K_2O (1.0), Na_2O (0.83), TiO_2 (0.81). This composition suggests potential sorption capacity the clay. The ratio [Al]/[Si] is about 1/3, which is optimal for ensuring required number and strength of surface active centers. In the examined sample, these centers are occupied mainly by Na^+ and K^+ ions; efficient exchange of the latter for heavy metal ions that form poorly soluble aluminosilicates in acid medium is quite possible. Therefore, the maximum sorption capacity of the clay may be estimated *a priori*. The overall concentration of Na_2O and K_2O in the clay sample is ~0.24 mmol/g, this value reflecting the maximum sorption capacity for doubly charged heavy metal cations. The major crystalline mineral component is montmorillonite while kaolinite and muscovite are concomitant phases. The presence of montmorillonite as the major phase (60%) is consistent with the [Al]/[Si] ratio equal to ~1/3 which reflects high potential ion-exchange capacity of the clay.

It is important to compare the experimental maximum sorption capacity for Ni^{2+} (Q = 0.11 mmol/g) – Table 1 – with the prediction made on the basis of the sorbent composition (0.24 mmol/g; see above). Presumably, accessible active surface of the sorbent contains approximately a half of the total amount of alkali metal ions, and just those active sites on the surface involved in ion exchange process determine the observed sorption capacity. As a result, the degree of extraction of Ni^{2+} is 95–97% at low Ni^{2+} concentration in solution ($c = 10^{-5}$–10^{-4} M).

Table 1 demonstrates that maximum sorption capacity Q = 0.11 mmol/g (0.22 mequiv/g) is consistent with the water monolayer capacity (a_m = 0.24 mmol/g). It looks as very specific and means that surface active sites accessible to water vapor are almost equally accessible to nickel ions in solution, which may be regarded as an indirect support for high degree of occupation of the sorbent surface by Ni^{2+} ions. In addition, estimation of the adsorption cross section of Ni^{2+} for Q = 0.11 mmol/g as $\omega = S_{sp}\, Q^{-1}\, N_A$ = 0.2 (nm^2) matches well with the hydrated nickel ion size.

Table 1. Sorption parameters of Ni^{2+} for three types of clay of the northwest region of Russia

	1	2	3
Q, mmol/g mequiv/g	0.11 0.22	0.25 0.50	0.31 0.62
ΔG, kJ/mol	−12.6	−13.5	−16.5
S_{sp}, m^2/g	14	30.5	36
Q, mequiv/m^2	0.016	0.016	0.017
a_m, mmol/m^2	0.017	0.017	0.017

In general, limits of sorption in relation to Ni^{2+} ions are in good agreement with the change in free energy ΔG. It is important to note that the sequence is determined by the totality of sorbent properties, including the important role of the specific surface S_{sp}. Revealing, in this regard, are the results of casting the limit sorption values Q and capacity of monolayer of adsorbed water a_m (table). Convergence of values obtained from normalization per m^2 testifies to a single nature and nearest sorption/ exchange surface activity of the investigated natural samples.

Finally, the results of potentiometric titration of an aqueous suspension of the sorbents are very representative. The initial pH value equal to 9.65 reflects equilibration between the glass electrode and alkaline surface of solid particles. The amount of acid necessary to adjust the suspension to a neutral pH value is 0.22 ± 0.03 mmol/g, which coincides with the maximum sorption capacity for Ni^{2+} (Q = 0.22 mequiv/g) and is close to the monolayer capacity of adsorbed water (a_m = 0.24 mmol/g). The observed agreement between independently determined parameters indicates high surface coverage of the clay sorbent with alkali metal ions and reflects their activity in both ion exchange and adsorption processes.

The main determinant of the set unity is the proximity of clay sorbents module-ratio, namely, Al/Si ~ 1/3.

ACKNOWLEDGMENTS

This work was financially supported by the Ministry of Education and Science of the Russian Federation in the framework of the basic part of the State contract.

REFERENCES

[1] Bhattacharyya K.G., Gupta S.S. *Advances in Colloid and Interface Science*, 140, (1), (2008) 114-131.

[2] Kumrič K.R., Duk č A.B., Trtič-Petrovič T.M., Vucelič N.S., Stojanovič Z., Grbovič-Novacovič J.D., Matovič L.L. *Industrial and Engineering Chemistry Research*, 52 (2013) 7930-7939.

[3] Ivanova E.S., Gavroskaya Yu.Yu. *Russian Journal of General Chemistry*, 84 (2), (2014) 181-184.

In: Applied Aspects of Nano-Physics …
Editors: Kirill Levine and Andrey Syrkov ISBN: 978-1-53614-708-7

Chapter 6

EXFOLIATED GRAPHITE FROM GRAPHITE NITRATE COINTERCALATION COMPOUNDS: PRODUCTION AND SOME APPLICATIONS

A. A. Voitash[1], V. Yu. Vishnevsky[1], Yu. V. Berestneva[1], E. V. Raksha[1,*], A. V. Muratov[1], A. B. Eresko[1], V. A. Glazunova[2], V. V. Burhovetskiy[2], G. K. Volkova[2] and M. V. Savoskin[1]

[1]L. M. Litvinenko Institute of Physical Organic and Coal Chemistry, Donetsk, Ukraine
[2]Donetsk Institute for Physics and Engineering named after A. A. Galkin, Donetsk, Ukraine

ABSTRACT

A triple graphite intercalation compound – graphite nitrate cointercalated with acetic acid and ethyl formate was used as a starting material for exfoliated graphite production. Carbon nanoparticles were obtained by sonication of the exfoliated graphite. Sorption capacity of the exfoliated graphite to benzene, engine oil and heavy oil was determined. Sorption of the Pd catalysts by exfoliated graphite from the cross-coupling reaction after target product extraction was studied.

Keywords: exfoliation of graphite, graphite intercalation compounds, carbon nanoparticles, sorption capacity, cross-coupling reaction

---

* Corresponding Author's Email: elenaraksha411@gmail.com.

INTRODUCTION

Exfoliation of graphite is usually performed by the intercalation of graphite to form a graphite intercalation compound (GIC), followed by heating of the GIC. Recently we obtained new graphite cointercalation compounds via additional intercalation of different organic substances into the graphite nitrate [1, 2]. It was revealed that variation in the organic cointercalants nature and composition allows one to control the stability/lability of the resulting compounds within a wide range [1, 2]. This paper presents the results on the exfoliated graphite production starting from triple graphite nitrate cointercalation compounds (GNCC) as well as some applications of the exfoliated graphite for carbon nanoparticles generation, and water purification from ecotoxicants.

RESULTS AND DISCUSSION

A triple graphite intercalation compound – graphite nitrate cointercalated with acetic acid and ethyl formate was obtained as a starting material for exfoliated graphite production. The overall scheme of the process involves graphite nitrate synthesis as a first stage followed by cointercalation with organic substances. The resulted GNCC dried at 20°C then undergoes thermal shock at 900°C to obtain the exfoliated graphite. The obtained GNCC was black-gray powder resembling initial graphite (Figure 1a) but some pre-expanded flakes (Figure 1b) were present in the sample even after drying at 20°C. Thus, this compound was expected to be perspective for nanoparticles [2] as well as exfoliated graphite production.

Samples of exfoliated graphite were sonicated (42kHz, 70W) in the ethyl alcohol for 1 hour to obtain carbon nanoparticles. The microstructure of obtained particles was investigated by transmission electron microscopy (TEM). It should be noted that started GNCC samples were compounds of the IV and II intercalation stages thus it was difficult to expect a complete graphite lattice exfoliation into monolayer graphene particles. TEM images of the obtained carbon nanoparticles are shown on Figure 2.

Exfoliated graphite has excellent sorption properties to some organic liquids, heavy oil or engine oil floating on the water surface. The sorption capacity of the exfoliated graphite to benzene, engine oil and heavy oil was determined. It was 71g, 43g, and 62g per 1g of the exfoliated graphite respectively.

As well, exfoliated graphite is promising sorbent for noble metals. The sorption of the Pd catalysts by obtained exfoliated graphite from the cross-coupling reaction after target product extraction was studied. We use 2-chloro-3-phenyl-pyridine synthesis (Figure 3) as an example of cross-coupling reaction that successfully applied in the large-scale pharmaceutical industry. SEM images of exfoliated graphite after Pd sorption are presented on Figure 3. Further investigation on the application of the obtained exfoliated graphite with Pd will be carried out.

(a) (b)

(c)

Figure 1. SEM images of initial graphite (a), graphite nitrate cointercalated with acetic acid and ethyl formate (b), and of exfoliated graphite (c) obtained from GNCC.

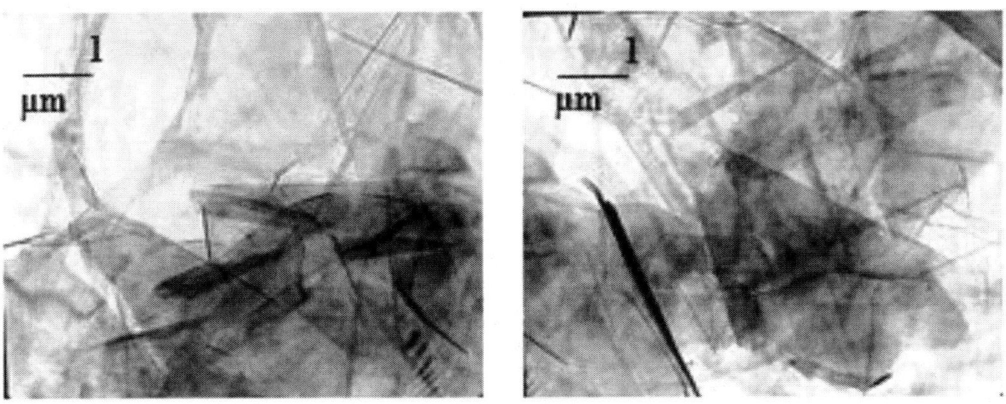

Figure 2. TEM images of carbon nanoparticles obtained by sonication of the exfoliated graphite.

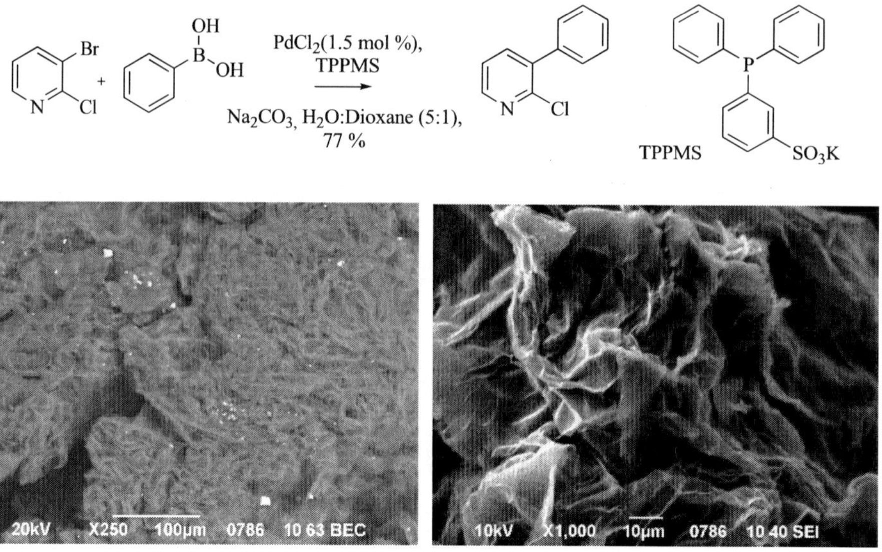

Figure 3. Scheme of the model reaction of 3-chloro-3-phenyl-pyridine synthesis and SEM images of exfoliated graphite after Pd sorption.

REFERENCES

[1] Savoskin, M. V.; Yaroshenko, A. P.; Whyman, G. E.; Mysyk, R. D. *J. Phys. Chem. Sol.,* 2006, vol. *67*, 1127 – 1131.

[2] Savoskin, M. V., Vdovichenko, A. N., Raksha, E. V., Berestneva, Yu. V., Vishnevskij, V. Yu., Padun, A. P., Verbenko, I. A., Yurasov Yu. I. Production of carbon nanoparticles by exfoliation of graphite nitrate and its co-intercalation compounds. In: *Proceedings of the 2016 International Conference on "Physics, Mechanics of New Materials and Their Applications".* Ed.: Parinov, I. A., Chang, Sh. H., Jani, M. A. - Nova Science Publishers, Inc., New York, 2017. – 794 p. – P. 115 - 120.

In: Applied Aspects of Nano-Physics ...
Editors: Kirill Levine and Andrey Syrkov

ISBN: 978-1-53614-708-7

Chapter 7

SOL-GEL SYNTHESIS, RESEARCH AND APPLICATION OF POWDERS – MULTIFERROICS WITH THE STRUCTURE OF A 'CORE-SHELL'

Olga A. Shilova[1,2,3,*]***, Tamara V. Khamova***[1]***,***
Gennady P. Kopitsa[1,4] ***and Maxim M. Sychev***[1,2]

[1]Institute of Silicate Chemistry of the Russian Academy of Sciences, Saint Petersburg, Russia
[2]Saint Petersburg State Technological Institute (Technical University), Saint Petersburg, Russia
[3]Electrotechnical University "LETI", Saint Petersburg, Russia
[4]National Research Centre Kurchatov Institute: Konstantinov Petersburg Nuclear Physics Institute, Gatchina, Russia

ABSTRACT

New composite powders of $BaTiO_3$ with a sol-gel-derived 'skin-layer' on their surface have been prepared. The data of SEM, XRD, X-ray spectral mapping, SAXS, powder neutron diffraction, small-angle scattering of polarized neutrons, low-temperature adsorption of N_2, magnetic characteristics of composite powders and dielectric parameters of prepared composites (polymer/powder) have been submitted.

---

[*] Corresponding Author's Email: olgashilova@bk.ru.

Keywords: powders, core-shell, powder mesostructure, fractal structure, organic-inorganic composite materials, magnetic properties, dielectric properties

Introduction

The multiferroics have both magnetic and ferroelectric properties, as well as a magnetoelectric effect due to the interaction of these components. This gives new opportunities in the design of electronic equipment.

Composite powders with a 'core-shell' structure: $BaTiO_3$ (core) /$CoFe_2O_4$ (shell) and more often in reverse order $CoFe_2O_4$ (core) /$BaTiO_3$ (shell) are successfully produced by co-precipitation and sol-gel processing [1, 2]. However, the mesostructure of the composites obtained, as well as the "synthesis conditions – structure – properties" dependencies, have not been adequately studied.

In the present work the results of studying the structure and properties of composite powders of the "ferroelectric (core). magnetic shell" type are considered.

Moreover these composite powders have been used as fillers in a polymer matrix. Composites: "polymer matrix /// multiferroics (filler)" are prospective for use in electronic engineering as film capacitors. A number of complementary modern research methods have been used.

Experimental Section

The 'skin-layers' have been formed during treatment of a commercial powder $BaTiO_3$ by tetraethoxisilane-derived sols containing compounds of cobalt and iron. The synthesized powders, so called multiferroics, were used as functional fillers for dielectric composites (polymer/filler) based on cyanoethyl ether of polyvinyl alcohol (CEEPVA).

The structure, composition and magnetic characteristics of composite powders and dielectric parameters of prepared composites (polymer/powder) have been investigated using the following methods: scanning electron microscopy (SEM), X-ray diffraction (XRD), X-ray spectral mapping, small-angle X-ray scattering (SAXS), powder neutron diffraction, small-angle scattering of polarized neutrons (SAPNS), low-temperature adsorption of N_2 and dynamic light scattering in the laser Doppler anemometer mode.

Results and Discussion

According to SEM, XRD and X-ray spectral analysis, the silica 'skin-layer' containing uniformly distributed nanoparticles of cobalt spinel ferrite ($CoFe_2O_4$) was formed on the surface of the powder. The specific surface of $BaTiO_3$ powder with a 'skin-layer' increased in ~ 17 times. This phenomenon can be associated with the formation on the powder surface of a more or less uniform layer, which changed the surface morphology. To confirm this hypothesis, X-ray spectral mapping of $BaTiO_3$ particles with a 'skin-layer' was carried out. It was shown that Si, Co and Fe are uniformly distributed over the surface of the $BaTiO_3$ particles. On the basis of the ζ-potential measurements the following assumption is made: the formation of 'skin-layer' takes place through the mechanism of electrostatic interactions of the oppositely charged powder particles and the sol particles ($\zeta_{BaTiO3} = -0.41$, and $\zeta_{sol} = +2.57$ at pH ~2).

The magnetic substructure of the powders with the structure of the "core ($BaTiO_3$)-shell ($SiO_2/CoFe_2O_4$)" was characterized by SAPNS. Estimations of the characteristic size R of magnetic-nuclear cross-correlations were obtained from the analysis of the patterns of magnetic-nuclear neutron scattering. The larger particles $CoFe_2O_4$ ~20-50 nm were observed using SEM. The magnetization curves for obtained materials were studied. The synthesized powders, so called multiferroics, were used as functional fillers for dielectric composites (polymer/filler) based on cyanoethyl ether of polyvinyl alcohol (CEEPVA).

The films based on the organic-inorganic composites (CEEPVA///$BaTiO_3$//SiO_2/ $CoFe_2O_4$) were formed under the influence of an external magnetic field. It has been found that these organic-inorganic composites based on the powders with a 'skin-layer', especially, which were dried under the influence of a magnetic field, have a higher permittivity ε as compared to the reference sample (CEEPVA///$BaTiO_3$) based on the initial powder [3].

Conclusion

A powder with a core-shell structure is prepared on the basis of commercial $BaTiO_3$ by treated with a doped TEOS-derived sol.

The 'shell' material consists of an amorphous silica matrix having a surface fractal structure formed by silica particles of 25-50 nm in size depending on the processing temperature. Particles of $CoFe_2O_4$ with a characteristic average size of 30-45 nm are statistically uniformly distributed in the silica matrix. At a temperature of 1000 °C, the silica matrix begins to crystallize.

The ferroelectric $BaTiO_3$ powder with a magnetic SiO_2 thin 'shell', containing nanoparticles of $CoFe_2O_4$, is magnetically hard material. It is used as filler in a polymer matrix made of cyanoethyl ether of polyvinyl alcohol to produce a composite material for film capacitors.

The use of a multiferroic composite powder of $BaTiO_3$ ('core') and $SiO_2/CoFe_2O_4$ ("shell") as filler as well as an external perpendicular directional magnetic field during the drying of a "multiferroic //polymer "composite greatly increases the dielectric constant ε of the composite material. These composites can be used as a material for embedded capacitors and other electronic devices, as well as sensors, shields of electromagnetic radiation, etc.

REFERENCES

[1] Schileo, G., Recent developments in ceramic multiferroic composites based on core/shell and other heterostructures obtained by sol–gel routes (2013) *Progr. Solid State Chem.* (41) 4: 87-98.

[2] Selvi, M. M., Manimuthu, P., Kumar, K. S., Venkateswaran, C., Magnetodielectric properties of $CoFe_2O_4$–$BaTiO_3$ core–shell nanocomposite (2014) *J. Magnetism Magnetic Mater.* 369: 155-161.

[3] Sychov, M., Nakanishi, Y., Vasina, E., Eruzin, A., Mjakin, S., Khamova, T., Shilova, O., Mimura, H. Core-shell approach to control acid-base properties of surface of dielectric and permittivity of its composite (2015) *Chem. Lett.* (44) 2: 197–199.

In: Applied Aspects of Nano-Physics …
Editors: Kirill Levine and Andrey Syrkov
ISBN: 978-1-53614-708-7

Chapter 8

THERMODYNAMIC MODELING OF THE HEATING OF FULLERENE C_{70} IN AN INERT ATMOSPHERE

Nick M. Barbin[1,2,3,*], Vasiliy P. Dan[1], Dmitriy I. Terentiev[1] and Sergey G. Alekseev[1]

[1]Ural Institute of State Fire Service of Emercom of Russia, Ekaterinburg, Russia
[2]Ural Agrarian State University, Ekaterinburg, Russia
[3]Ural Federal University, Ekaterinburg, Russia

ABSTRACT

Thermodynamic modeling of the behavior of fullerenes C70 by heating in argon at atmospheric pressure.

Keywords: warming, carbon nanomaterials, fullerene, thermodynamic modeling

INTRODUCTION

The study of the thermal stability of condensed fullerene structures is of interest for a range of practical applications of fullerenes, and for development of technologies for their production and purification [1]. Various sources considered temperature stability of fullerenes, however, due to discrepancies in the results a clear conclusion cannot be done.

---

* Corresponding Author's Email: nmbarbin@mail.ru.

This paper studied the behavior of C70 fullerenes under heating in argon at atmospheric pressure.

The research was conducted using thermodynamic modeling.

THE METHOD OF CALCULATION

Thermodynamic modeling consists of the thermodynamic analysis of the equilibrium state of the system as a whole. One of the most developed and effective programs that implement such thermodynamic calculations, is the software package TERRA [2].

Calculations of the composition of the phases and characteristics of the equilibrium are performed using reference databases [3-6]. Thermodynamic modeling was successfully used in [7, 8].

RESULTS AND DISCUSSION

Computer experiment allows to determine the phase distribution of carbon in quasi-equilibrium system C_{70}-Ar on the whole considered temperature interval. Calculation methods defined temperature interval of thermal stability of the fullerene C_{70}.

In the temperature range from 473 K to 2273 K the content of condensed C_{70} in the system C_{70}-Ar remains unchanged at 0,238 mol/kg In the temperature range from 2373 K to 3273 K the content of condensed C_{70} is reduced to 0,233 mol/kg. This is due to the beginning of the process of sublimation of the condensed component of the system and the emergence of gas C, C_2, C_3, C_4, C_5. In the interval from 3373 K to 3873 K content of condensed C_{70} decreases to 0,077 mol/kg At a temperature of 3773 K observed inflection point. Further increase of temperature resulted a linear decrease in the content of condensed C_{70}.

In the interval of temperatures from 2273 K to 3973 K the content of gaseous C and C_2 increases to 0,855 mol/kg 0,825 mol/kg At a temperature of 3973 K we observe inflection point of curves of gaseous C and C_2 from the temperature. This is due to the rapid decrease in the content of condensed C_{70}, as well as a decrease in the content of gaseous C_3, C_4, C_5 in the system C_{70}-Ar. The increase of temperature from 3973 K to 4273 K leads to a decrease in the intensity increase of the content of vapor and C_2; the maximum value is achieved at a temperature 4273 K – 2.178 mol/kg and 1,391 mol/kg, respectively.

Vapor content C_3 in the temperature range from 2273 K to 3873 K increases and reaches 2,887 mol/kg At a temperature 3873 K observed inflection point. In the temperature interval from 3873 K to 3973 K intensity of the vapor content increase for C_3

is reduced; the content of gaseous C_3 reaches the maximum value, 4,194 mol/kg. Increasing of the temperature to 4273 K resulted to a linear decrease in the content of gaseous C_3 to 3,650 mol/kg.

The content of C_4 and C_5 vapors in the temperature range from 2273 K to 3973 K increases and reaches 0,072 mol/kg 0,256 mol/kg at temperatures 3973 K in the graphs of dependence of the concentrations of the gaseous C_4 and C_5 is observed in the inflection point caused by a decrease in the content of condensed C_{70} and increased concentrations of gaseous C and C_2. In the temperature range from 3973 K to 4273 K content of gaseous C_4 and C_5 decreases linearly to 0,054 mol/kg 0,105 mol/kg, respectively.

CONCLUSION

As thermal stability was taken the temperature at which 10% of the total number of condensed fullerenes C_{70} has undergone dissociation.

According to the results of a computer experiment, fullerene C_{70} retains thermal stability up to a temperature 3673 K.

REFERENCES

[1] Borodin V.I., Trukhacheva V.A. *Thermal stability of fullerenes. Modern high technologies*. – 2004. – No. 1. – P. 82-84.

[2] Vatolin N.A. Moiseev G.K., Trusov B.G. Thermodynamic modeling in high temperature systems. M.: *Metallurgy*, 1994. 352.

[3] Moiseev G.K., Vatolin N.A. Computer simulation of the formation of various condensed forms of carbon. *Journal of Physical Chemistry*. – 2002. – T. 76. – No. 8. – Pp. 1366-1370.

[4] Moiseev G.K., Vatolin N.A. Evaluation of the thermodynamic properties of a number of condensed carbon compounds. *Journal of Physical Chemistry*. – 2002. – T. 76. – No. 3. – S. 424-428.

[5] Moiseev G.K., Vatolin N.A. Thermodynamic properties of some gaseous fullerenes. *Journal of Physical Chemistry*. – 2002. – T. 76. – No. 2. – P. 217-220.

[6] Moiseev G.K., Vatolin N.A. An estimate of the standard enthalpy of formation (sea) of metastable condensed small coassociation of carbon and certain metals. *Doklady of the Russian Academy of Sciences*. – 2003. – T. 392. – No. 5. – P. 653-656.

[7] Barbin Nick M., Dan Vasiliy P., Terentiev Dmitriy I., Alekseev Sergey G. Modeling the Behavior of Carbon Nanoparticles C44 When Heated in an Argon

Atmosphere. Computer Experiment. *Smart Nanocomposites* 2016. Volume 7 Issue 1.

[8] Kolbin T.S., Terentiev D.I., Barbin N.M. Alexeev S.G. Behaviour of Eu, Pu, Am at burning radioactive graphite in oxygen atmosphere at different pressures. *Technosphere Safety*. – 2013. – No. 1. S. 30-36.

In: Applied Aspects of Nano-Physics ...
Editors: Kirill Levine and Andrey Syrkov
ISBN: 978-1-53614-708-7

Chapter 9

THEORETICAL INVESTIGATION OF THE ATOMIC AND ELECTRONIC STRUCTURE OF SUBPORPHYRINAZIN AND SUBPHTHALOCYANINE

A. S. Kholtobina*, E. A. Kovaleva and P. O. Krasnov
Department of Physical and Inorganic Chemistry
Siberian Federal University, Krasnoyarsk, Russia

ABSTRACT

Atomic and electronic structures of subporphyrinazin and subphthalocyanine were investigated by the density functional theory method using a planar wave basis set and PAW approximation.

Keywords: subporphyrinazin, subphthalocyanine, DFT, density of states

In the past decade, SubPcs have found significant application potential in various advanced technology fields such as nonlinear optical (NLO) functional dyes, sensors, light-emitting diodes (OLEDs) [1]. Subporphyrazines (SubPzs) become an attractive class of chromophore molecules because of their intense Q band absorption and fluorescence in the visible region [2].

* Corresponding Author's Email: akholtobina-a15@stud.sfu-kras.ru.

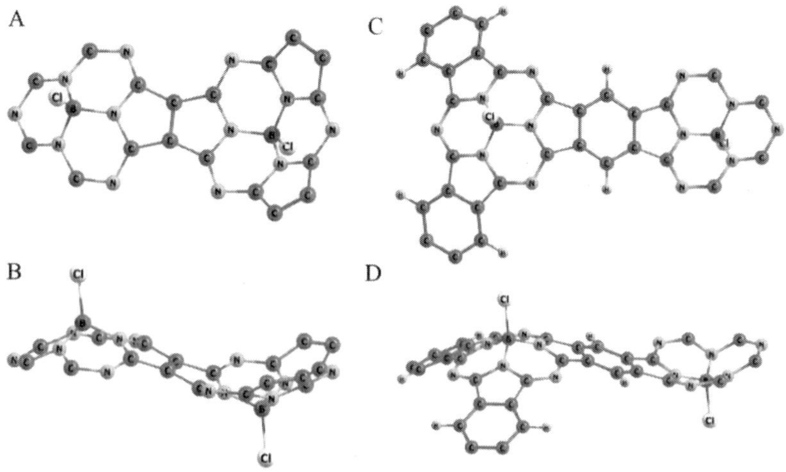

Figure 1. Geometries of Subporphirinazin (top and side views A, B) and subphthalocyanine (top and side views C, D).

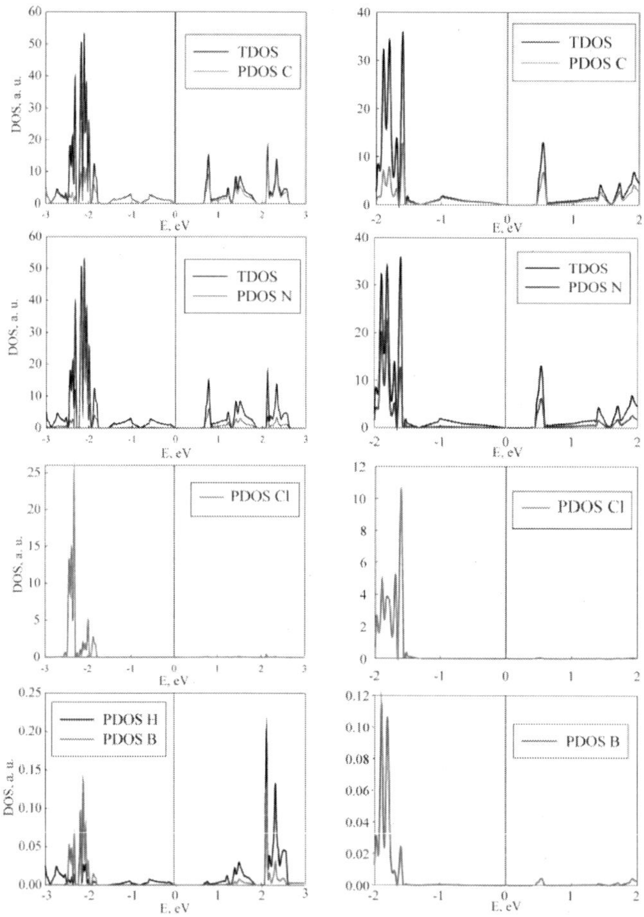

Figure 2. Total and partial densities of states for subporphirinazin (left side) and subphthalocyanine (right side).

However, in the absence of theoretical works describing their electronic properties, it is interesting to investigate their electronic structure, particularly densities of states using DFT.

Quantum chemical simulation of the objects under study was performed by the DFT method using the VASP program package. All calculations given in this work were carried out for 6*6*1 k-point mesh. The densities of states (DOS, total – TDOS, partial – PDOS) were obtained by the Gaussian smearing with a peak half-width of 0.02 eV.

According to the densities of states, both subporphyrinazin and subphthalocyanine are semiconductors. Analyzing PDOS, it should be mentioned that HOMO is mostly contributed by the carbon atoms while LUMO has the supply from nitrogen, boron and hydrogen atoms in the case of subphthalocyanine.

REFERENCES

[1] Wang, C., Chen, X., Qi, D., Bi, S., Jiang, J., (2017). Unraveling the formation mechanism of subphthalocyanine. Density functional theory studies. *Inorganic Chemistry Communications,* in press.

[2] Liang, X., Shimizu S., Kobayashi N., (2014) Sizeable red-shift of absorption and fluorescence of subporphyrazine induced by peripheral push and pull substitution. *Chem. Commun.*, 50, 13781.

In: Applied Aspects of Nano-Physics …
Editors: Kirill Levine and Andrey Syrkov
ISBN: 978-1-53614-708-7

Chapter 10

THE INVESTIGATION OF SIGNAL MULTIPLICATION MIXER LINEARITY AND AMPLIFIER SCHEMES

Yu. A. Nikitin and V. A. Yurova*
The Bonch-Bruevich Saint-Petersburg State University
of Telecommunications, Saint-Petersburg, Russia

ABSTRACT

In this chapter, we obtained the basic spectral and temporal characteristics of the output signal of the frequency amplifiers and mixers on the basis of the constructed schemes of the spectrum analyzers. We present the results of a comparative analysis of the linearity of the signals multiplication by using a generic approach and the same parameters of the active elements.

Keywords: bipolar transistor, frequency multipliers, spectrum analyzer, spectral characteristics, temporal characteristics

INTRODUCTION

The signal mixers and amplifiers are widely used in different devices of modern electronics, in its manufacturing and the investigation of nanostructures. Therefore, one of the major tasks in the manufacture of electronic equipment is to provide the high accuracy of the measurement of the parameters of various materials in small quantities

* Corresponding Author's Email: va-yurova@mail.ru.

and in its multi-component combinations. It means that such devices as spectrum analyzers make the strict demands on the linearity of the multiplication of the measurement signals at the circuits of amplifiers and frequency mixers, i.e., the level of nonlinearity of the active devices – semiconductor transistors, diodes etc. Generally, the different approaches, evaluation indicators and the mathematics are used in the investigations, devoted to the analysis parameters and operation modes of amplifiers and frequency converters. It is difficult to compare the parameters of the considering electrical circuits. So, the aims of our research are obtaining spectral and temporal characteristics of the circuits of amplifiers and frequency mixers based on transistors, and a comparative analysis of the linearity of the multiplication, using a common approach and the same active element parameters.

OBJECT OF RESEARCH

The objects of research were chosen electric circuits based on a transistor cascade and cascade current mirror (Figure 1). In the second case, two types of connections of bipolar transistors in the circuit-series and parallel series were considered. These electric circuits are widely used as individual cascades, and are the basis for designing more complex circuits of frequency mixers, which are widely used in electronics and measuring technology of the manufacturing and the laboratory investigations of nanoelectronic devices and materials of the nanotechnology [1-4].

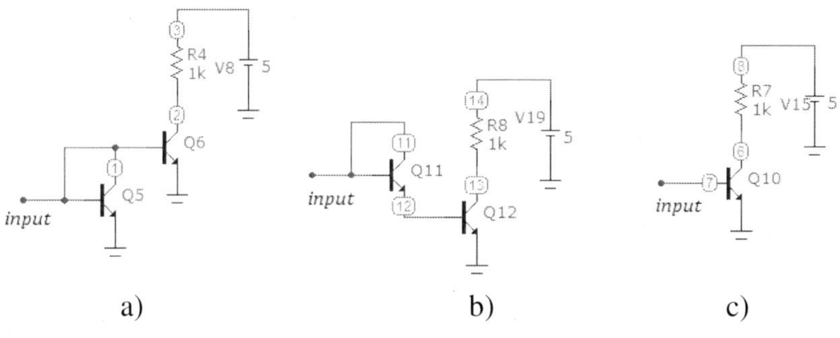

Figure 1. The diagrams of the electric circuits of cascades based on the current mirror with series (*a*) and parallel (*b*) transistors connection and the transistor cascade (*c*).

METHOD OF RESEARCH

To obtain the spectral and temporal characteristics of the output signal and to make the comparative analysis of the linearity's parameters of the cascade electric circuits, the computer program for the simulation of electrical circuits MicroCap 9.0 was used. It makes it possible to carry out spectral research of the different structures of the electric circuits over a wide dynamic range, applying a built-in database of the electronic components, which are used in modern electronics. The criterion of linearity of the multiple signals in mixers and amplifiers is the level of combinational components higher than the second level, because the combinational component of the second level $f_{OUT} = f_{IN} \pm f_{HET}$ is the desired output signal. Here, the frequency f_{HET} is the parameter of the heterodyne. If the electric circuits include a power source (like on Figure 1), it work as an amplifier; if no, then it works as a signal mixer.

RESULTS AND DISCUSSION

In the research we calculated parameters of the cascade electric circuits, which allow for the specified amplitude of the output signal and the transfer coefficient.

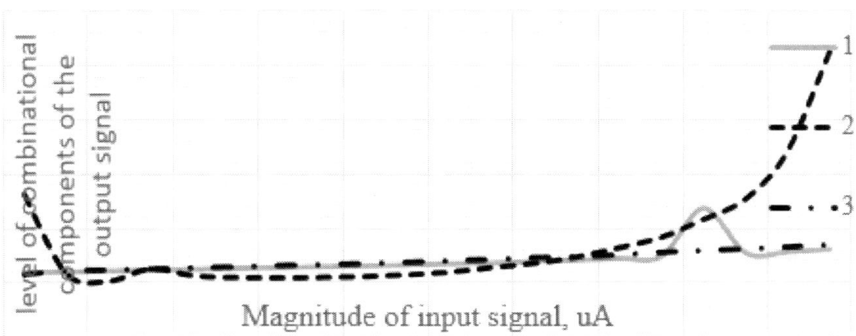

Figure 2. The dependence of third level of combinational components of the output signal of the electric circuits of cascades based on the transistor cascade (*1*) and the current mirror with parallel (*2*) and series (*3*) transistors connection.

Computer simulations of the frequency mixers and amplifier schemes were made of common criteria, using the common computer simulation environment. Figure 2 shows the obtained dependence of the third level of combinational components of the output signal on the magnitude of the input signal from the current source of the amplifiers based on a transistor cascade and cascade current mirror. When schemes are used as signal mixers, the dependence of second level combinational components of the output signal is investigated.

CONCLUSION

The main spectral and temporal characteristics of the output signal for electric circuits of the frequency mixers and amplifiers based on the transistor cascade and the current mirror cascade with series and parallel transistor connections were obtained. These electric circuits are widely used as the frequency amplifiers and mixers in different devices, such as spectrum analyzers, and as the basis to design more complex electrical circuits of frequency changers, such as balanced and circular mixers. The analysis of results is shown so the base cell mixer based on the current mirror provides a lower level of combinational components. It's important to develop high-quality equipment for measuring different parameters of materials and nanoscale structures and manufacturing of modern electronic devices. Also, it was found that the circuit of the current mirror with a serial connection of bipolar transistors is characterized by the higher linearity when the current generator is connected as the signal source. In the case of using a voltage generator as a signal source, the circuit of the current mirror with the parallel connection of bipolar transistors is characterized by the higher linearity.

REFERENCES

[1] Polyakov, A. E., Strygin, L. V. 2-Tone IP2 and IP 3 Measurement Technique. *Trudy MIPT,* 2012, №2. – P. 54 - 63.

[2] Gilbert, B. The Micromixer: a highly linear variant of the Gilbert mixer using a bisymmetric Class-AB input stage e. *IEEE Journal of Solid-State Circuits,* 1997. Vol. 32, N. 9. – P. 1412 - 1423.

[3] Fomin, N. N., Buga, N., Golovin, O. V. *Radio receivers: textbook for universities. - M: Techbook,* 2007. - 520 p.

[4] Jones, M. H. *A practical introduction to electronic circuits.* Third edition. Cambridge University Press, 1996. - 548 p.

In: Applied Aspects of Nano-Physics …
Editors: Kirill Levine and Andrey Syrkov
ISBN: 978-1-53614-708-7

Chapter 11

CONTROL OF CURRENT AND VOLTAGE OSCILLATIONS IN A SHORT DC DISCHARGE MAKING USE OF EXTERNAL AUXILIARY ELECTRODE

A. S. Mustafaev[1], *A. Y. Grabovskiy*[1],
V. I. Demidov[2,*], *I. Kaganovich*[3] *and M. E. Koepke*[2]
[1]Saint Petersburg Mining University, Saint Petersburg, Russia
[2]West Virginia University, Morgantown, West Virginia, US
[3]Princeton Plasma Physics Laboratory, Princeton, New Jersey, US

ABSTRACT

A dc discharge with a hot cathode is subject to current and voltage plasma oscillations, which have deleterious effects on its operation. In this paper, we demonstrate how it is possible to achieve reliable suppression of the discharge oscillations by installing an auxiliary electrode, placed outside of anode. By collecting a modest current through a small opening in anode, we show that the discharge becomes stable, in a certain pressure range. This method of avoiding current oscillations can be used for wide range of applications.

Keywords: plasma chemistry, dc discharge, electron distribution function, plasma instabilities negative resistance, heated cathode

Gas-discharge plasma devices are widely used in plasma applications [1]. A better understanding of the physics behind these devices has been developed over the last few

* Corresponding Author's Email: Vladimir.Demidov@mail.wvu.edu.

decades is made possible through development of new diagnostic techniques [2] and has allowed for the creation of discharge plasmas with more controllable parameters, including modification of the charged particle densities, temperatures (average energies), and electron energy distribution functions (EEDF). However, improving the performance of these new plasma tools is still a significant challenge for plasma engineering. Exploitation of nonlocal plasma properties allows additional dimensions and flexibility in adjusting plasma parameters. A remarkable property of such plasmas is that changing conditions in one place may lead to unexpected changes far away in another part of the plasma. One example of such a device with nonlocal plasma properties is a short dc discharge (several millimeters in length at the pressure of a few Torr, 10–100 μm for atmospheric pressure).

It is known [3] that dc electric discharges can be unstable with respect to excitation of various types of oscillations and instabilities. Oscillations are excited due to ionization plasma instability related to the falling volt-ampere characteristic of the discharge or part of it. These oscillations are affected by external electric circuits. There are many regimes of oscillations, which can be identified. While plasma instabilities, in principle, can be harnessed for some purposes, for example, for the generation of voltage oscillations, they are harmful for many other applications, for instance, for the development of current and voltage stabilization devices, and, in this case, they should be suppressed. In this paper, we demonstrate how it is possible to achieve reliable suppression of the discharge oscillations by placement of the external auxiliary electrode outside the discharge and subtraction of the anode current through a small opening in the anode.

In the figure the typical current-voltage experimental characteristics of the cathode-anode gap are presented. Curves 1, 2 and 3 correspond to the case without collection of the current by the auxiliary electrode. It is possible to see that in the diode regime, the discharge has a positive discharge differential resistance for the pressures below 1 Torr (curve 1). Increasing the gas pressure to 1 Torr leads to transformation of the discharge differential resistance to the slightly negative one (curve 2). Further increase of the gas pressure yields higher negative discharge differential resistance (curve 3). The presence of the negative resistance may lead to plasma instabilities and oscillations of the discharge voltage and current.

The collection of the anode current to the auxiliary electrode leads to a transformation of the discharge differential resistance from the negative to the positive one, as evident from curves 4, 5, and 6. Curves 4 and 5 were obtained for the helium gas pressure of 1 Torr and curve 6 is for 4 Torr. Each curve was obtained with fixed auxiliary electrode current. The experimental data show that an increase in the current drawn to the auxiliary electrode makes discharge differential resistance even more positive (compare curves 4 and 5). In all cases with the positive differential resistance, the oscillations are suppressed and are practically absent.

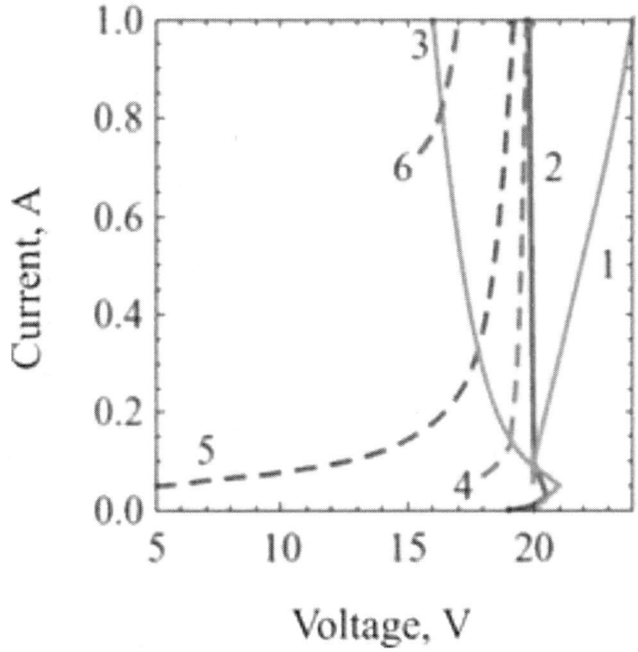

Figure 1. The discharge current-voltage characteristics for helium gas pressure of 0.6 Torr (1), 1 Torr (2), and 4 Torr (3). The diode regime is shown by solid curves (no current to the auxiliary electrode). Regimes with the auxiliary electrode are shown by the dashed curves for helium gas pressure of 1 Torr and different currents to the auxiliary electrode (0.1 A (4) and 0.4 A (5)). Curve (6) is for helium gas pressure of 4 Torr and current to the auxiliary electrode of 0.1 A.

REFERENCES

[1] Mustafaev, A.S., Grabovskiy, A.Y. *High Temperature*. 2017. Vol.55. No1. P.20-26.

[2] Mustafaev, A.S., Soukhomlinov, V.S. *Journal of Mining Institute*. 2016. Vol. 222. P.864-868.

[3] Mustafaev, A.S. *Technical Physics*. 2001. Vol.46. №4. P.472-483.

In: Applied Aspects of Nano-Physics …
Editors: Kirill Levine and Andrey Syrkov
ISBN: 978-1-53614-708-7

Chapter 12

EFFECTS OF GRAPHITE INTERCALATION WITH CESIUM IN A THERMIONIC ENERGY CONVERTER

A. S. Mustafaev* and V. S. Sukhomlinov
Saint Petersburg Mining University, Saint Petersburg, Russia

ABSTRACT

The thermionic energy converter (TEC) with inter-electrode low-temperature plasma and cesium vapor dynamic flow in the inter-electrode gap (IEG) demonstrated an increase in efficiency of up to 20%. This was mainly reached due to the decrease in the effective electron work function to an anomalously low value of the order of 1 eV from a perforated nickel collector covered with nanosized graphite flakes under the converter working conditions. SEM X-ray microanalysis of the collector surface layers was performed, and the model of the effects proposed. In a such kind of collector of TEC with a nanosized graphite coating, multilayered intercalation of Cs atoms in the graphite/graphene surface layers was found during work in cesium vapors at a pressure of ~1–10 Torr. This probably explains the anomalously low electron work function and the high efficiency of TEC.

Keywords: thermionic energy converter, electron work function, nickel collector, nanosized graphite, elemental microanalysis, graphite intercalation by cesium atoms

---

* Corresponding Author's Email: alexmustafaev@yandex.ru.

Composite carbon nanostructures with a low work function are of considerable interest for numerous applications ranging from electron emission devices to ion-battery technologies. It is known that adding hydrogen atoms or alkali metals to the surface of carbon structures can lead to remarkable reduction of the surface work function. These surface effects can be used in new thermionic energy converters (TEC) of the thermal power into electric.

Studies on TEC performed at the Institute for Physics and Power Engineering and aimed at seeking methods for drastically raising the efficiency (several-fold) and operating life (by approximately an order of magnitude) relative to the previously achieved level (~10% and ~1 year, respectively) in TOPAZ and Yenisei space nuclear power units [1] showed that it is possible to achieve high efficiency of converters with electrodes of advanced materials [2-4].

As is well known, the electromotive force developed by TEC is mainly determined by the electron work function (EWF) of the emitter material, and the output voltage losses depend on the EWF of the collector material and plasma voltage losses in the interelectrode gap (IEG). The efficiency of TEC can be increased mainly by decreasing the collector EFW. Earlier [5, 6], high efficiency of laboratory TEC was demonstrated experimentally by continuously circulating cesium vapors through a nickel collector with many holes with a diameter of 0.1 mm. The effect was observed if the collector surface was coated with graphite using an aquadag suspension, and not observed on a metal surface. The experiments performed in [5, 6] were repeated independently on an improved laboratory TEC with a perforated collector with a coating of nanosized graphite flakes.

As TEC electrodes, we used an emitter of vacuum-melted polycrystalline Mo with a diameter of 14 mm and a thickness of 11 mm with a Pt (~3 μm) coating and a collector (with a diameter of 8 mm) of laser-perforated nickel foil with a thickness of 0.2 mm (121 holes with a diameter of 0.1 mm on a square of 4×4 mm). The emitter was heated by electron bombardment; T_E = 1350 K, T_C = 750 K.

The panoramic SEM images of the collector surface show concentric rings around each hole in the collector (figure). They evidently suggest that diffusion occurs on the collector surface from the boundary of the hole.

Then the distribution of elements between the neighboring holes was measured in different regions of the collector. Spectrum 12 from the region of the concentric diffusion ring in the figure showed Ni, Mo, C, Cs, and O lines. During the operation of TEC, molybdenum evaporated from the emitter and fell on the collector. In spectrum 7, there were no Cs and O X-ray lines outside the limits of the concentric diffusion ring in the figure. Nearly equal partial concentrations of Cs and O atoms were observed at different points of the diffusion rings around the collector holes, but not observed outside the holes. This may be explained by diffusion (intercalation) of these atoms from the hole boundaries into surrounding multistoreyed (with a height of ~1 μm) layers of nanosized graphite-graphene. The observed Ni concentration inside the rings is lower than outside because Cs and O intercalated into multilayered graphite-graphene screened the lower-lying Ni atoms of the substrate from fast electrons.

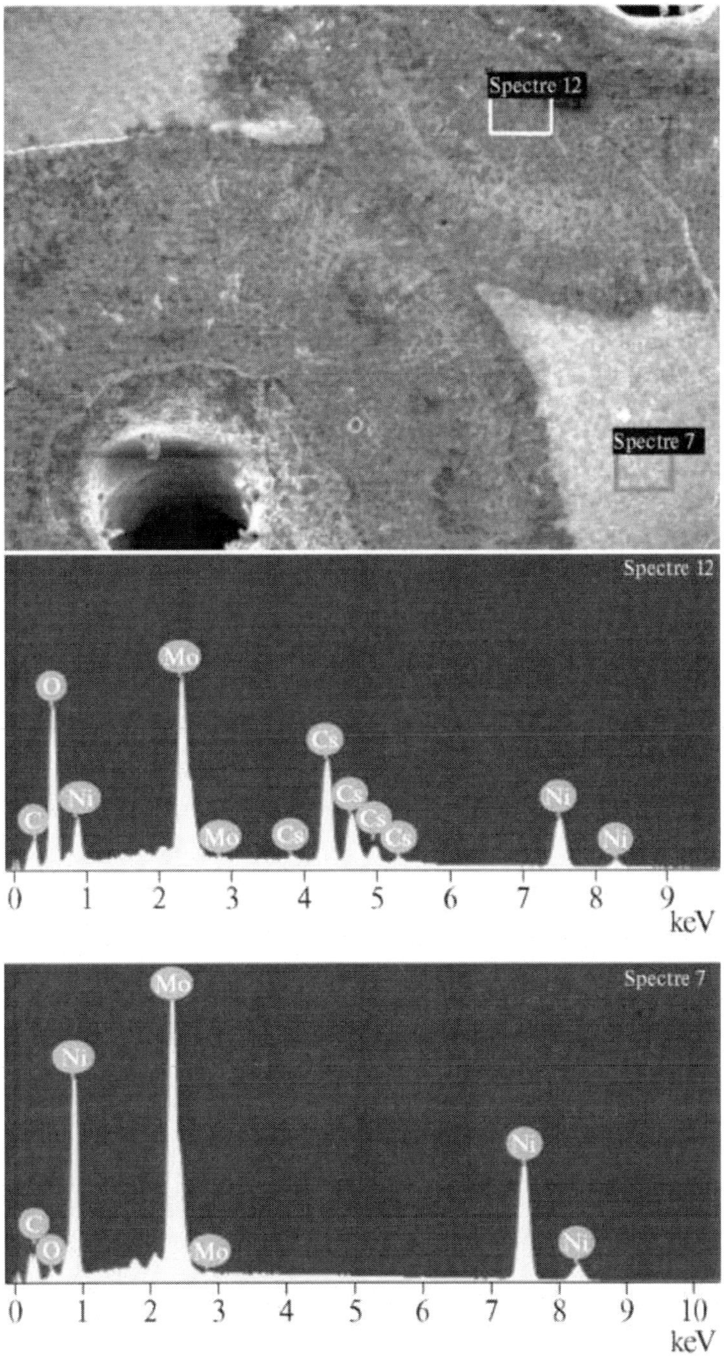

Figure 1. SEM image and X-ray fluorescence spectra of the surface of a perforated nickel collector coated with nanosized graphite after prolonged exposure to Cs vapors.

REFERENCES

[1] G. M. Gryaznov and V. Ya. Pupko, *Priroda,* (Moscow, Russ. Fed.), No. 10, 29 (1991).

[2] V. I. Yarygin, *J. Clust. Sci.,* 23, 77 (2012).

[3] V. I.Yarygin, A. S. Mustafaev, *Russian Journal of Physical Chemistry B,* V.9, No. 4, p. 546 (2015).

[4] A. S. Mustafaev et al., *Russian Journal of Physical Chemistry B*, V.36, No.2, p.94 (2017).

[5] L. Holmlid, in *Proceedings of the Thermionic Energy Conversion Specialist Conference,* Goteborg, Sweden, 1993, p. 47.

[6] R. Svensson, L. Holmlid and E. Kennel, in *Proceedings of the Thermionic Energy Conversion Specialist Conference*, Goteborg, Sweden, 1993, p. 93.

In: Applied Aspects of Nano-Physics ...
Editors: Kirill Levine and Andrey Syrkov

ISBN: 978-1-53614-708-7

Chapter 13

INVESTIGATION AND FORMATION OF A SORPTION GAS SENSOR BASED ON CARBON NANOSTRUCTURES WITH COMPLEX RELIEF

A. A. Rezvan *and V. S. Klimin*
South Federal University, Taganrog, Russia

ABSTRACT

In this chapter, design and a laboratory-technological instruction for the manufacture of a sorption gas sensor with a sensitive element based on an array of vertically oriented carbon nanostructures have been developed. The essence of this technology lies in the production of graphene films by the method of plasma-chemical deposition from the gas phase on the surface of vertically-oriented nickel nanostructures.

Keywords: nanotechnology, graphene, carbon nanostructures, sorption sensors of gas, plasma chemical vapor deposition (PECVD)

INTRODUCTION

At present, various researchers associate further prospects for the development of nanotechnologies with carbon-based structures. Such close attention and wide interest of science carbon nanomaterials were caused primarily by the discovery of fullerenes, carbon nanotubes and graphene, which have unique physical and mechanical properties.

* Corresponding Author's Email: arezvan@sfedu.ru.

Thus, due to its box structure, resistance to ionizing radiation and a hexagonal two-dimensional monatomic crystal lattice, which makes it possible to determine small concentrations of adsorbed gas, carbon nanostructures are one of the promising materials of micro- and nanoelectronics. In this connection, development and creation of portable gas sensors on their basis is actual.

EXPERIMENTAL SECTION

In the course of this work an experimental study of forming a sorption gas sensor using vertically oriented carbon nanostructures method was carried out. Experimantal layout is a small insulating substrate on which measuring electrodes, a sensor element and a heater are located on opposite sides of the substrate. The heater is necessary for degassing, restoring the sensitive element and maintaining a constant temperature of the processes occurring on the surface of the semiconductor during gas chemisorption.

So, initially, Experimental layouts were chemically purified subtrates Si(100), on which silicon dioxide Si3N4 films were synthesized by the method of plasma chemical vapor deposition(PECVD). Also, using the magnetron sputtering method, a catalytic Ni layer was formed on this structure. On the obtained structure, nickel nanowires were formed by a combination of methods of focussed ion beams and plasma chemical etch. Carbon nanostructures on the profiled surface of this experimental model were obtained by plasma-chemical deposition from the gas phase, the carbon-containing gas was acetylene C_2H_2. To the finished model, over the sensitive layer, electrical contacts connected to the signal processing device were connected.

RESULTS AND DISCUSSION

As a result of the experimental studies, it was determined that the made model of the resistive gas sensor based on the sensing element consisting of a carbon nanosystem is susceptible to acceptor molecules of gas NO_2 and donor molecules of gases NH_3, CO, H_2O vapor with different degrees of sensitivity depending on the concentration n.

REFERENCES

Ageev O.A., Klimin V.S., Solodovnik M.S., Eskov A.V., Krasnoborodko S.Y. The study of influence of the gas flow rate to etched layer thickness, and roughness of the anisotropy field of gallium arsenide is etched in the plasma chemical etching process. *Journal of Physics: Conference Series*, 2016, T. 741, № 1. p. 012178.

Klimin V.S., Solodovnik M.S., Smirnov V.A., Eskov A.V., Tominov R.V., Ageev O.A. A study of the vertical walls and the surface roughness GaAs after the operation in the combined plasma etching. *Proceedings of SPIE - The International Society for Optical Engineering "International Conference on Micro- and Nano-Electronics 2016*", 2016, p. 102241Z.

Ilina M.V., Blinov Y.F., Ilin O.I., Klimin V.S., Ageev O.A. Resistive switching of vertically aligned carbon nanotube by a compressive strain. *Proceedings of SPIE - The International Society for Optical Engineering "International Conference on Micro- and Nano-Electronics 2016,"* 2016, p. 102240U.

In: Applied Aspects of Nano-Physics …
Editors: Kirill Levine and Andrey Syrkov

ISBN: 978-1-53614-708-7

Chapter 14

INVESTIGATION OF THE INFLUENCE OF THE MODIFICATION PROCESSES OF NANOSTRUCTURES MICRORELIEF BASED ON SILICON DIOXIDE FILMS BY THE METHOD OF PLASMA-CHEMICAL ETCHING

V. S. Klimin**, A. A. Rezvan and I. N. Kots
South Federal University, Taganrog, Russia

ABSTRACT

In this chapter, the use of a plasma chemical etch method was examined for implementation of profiling processes of surface layers of structures based on silicon dioxide SiO_2, which can be used for further epitaxial growth. The essence of the technology is a rational sequence of methods of focused ion beams and plasma chemical etch(PCE). Preproduction models were formed with unique surface relief, which depends on doses of embedded ions.

Also, the range of dose values of the Ga + ion beam for silicon doping was determined.

Keywords: silicon dioxide, local ion doping, plasma chemistry, focused ion beam, nanotechnology, semiconductor structures

* Corresponding Author's Email: arezvan@sfedu.ru.

INTRODUCTION

The future micro- and nanoelectronics plasma chemical etch or PCE of semiconductor structures is one of the main methods for synthesizing nanoscale structures with sufficiently low linear dimensions less than 100 nm. This method for the formation of etching process uses chemical active particles, which are generated in a low-temperature plasma of a high-frequency (HF) discharge. Because of this, it is possible to control changes in the concentrations of electrons and ions in the discharge, which can significantly affect the demand for the characteristics of the etching process.

So, PCE is characterized by high resolution, selectivity and anisotropy of etching. In connection with this, the choice of the method of chemical etching from the high-frequency (HF) plasma discharge for the surface modification processes is actual.

EXPERIMENTAL SECTION

In this paper an experimental study of the effect of parameters of the processes of modifying and surface treatment of structures based on silicon dioxide films by the method of plasma chemical etching. Experimental layouts were chemically purified subtrates Si(100), on which silicon dioxide SiO2 films were synthesized by the method of plasma chemical vapor deposition (PECVD).

Further local doping of silicon dioxide regions with gallium ions Ga^+ was carried out.on the surface of obtained structures by the method of focused ion beams. This process was conducted on a scanning electron microscope “Nova NanoLab 600” with ion column and scanning probe microscope (SPM), with the help of which the samples were scanned. The order of the beam movement was determined by the required sequence of coordinates in the template file. The beam propagation time varies from 100 to 1000 ms. This pattern file is executed with apertures with a beam current of 1, 10, 30 pA with a beam diameter of 7, 12, 16 nm, respectively, and an ion beam energy of 30 keV in the absence of ion beam overlap at neighboring points of impact.

There selective parameters of plasmachemical etching processes for each mask were used for the formation of nanoscale structures based on silicon dioxide films. Experimental layouts were subjected to PCE in a high-frequency inductively-coupled plasma in a " STE ICP E68" installation in several stages.

The first stage was the etching of silicon dioxide films of experimental samples, the working gas mixture of the process being SF_6. The second stage was final, the structures were etched in a gas mixture of BCl3 to remove the remaining Ga particles.

The determining parameters of the silicon dioxide PCE were chosen so that the chemical etching mechanism predominated over physical spraying. With the selectivity condition, volumetric nanoscale structures were obtained with the following parameters of PCE: RF power supplied to the source inductor, 200 W; capacity of the capacitive pair supplied to the substrate, 10 W; the pressure in the chamber is 1.33 Pa; the bias voltage is 40 V, the working gas flow is N_{SF6} = 15 cm^3 / s. To obtain the etching depth of silicon 9.73 and 252.1 nm with these parameters, the etching time is 30 and 120 s, respectively.

The bias voltage is necessary to accelerate the etching process, but it was chosen in such a way that the etching rate of silicon was significantly greater than the etching rate of the gallium mask. Further investigations of the surface structure were carried out by scanning electron microscopy and probe nanotechnology.

Results and Discussion

After the completion of a number of experimental studies, the dependence of the geometric parameters of the etched layer on the etching time was obtained, and an analysis was made of the rates of the etching process for various gas flows. In addition, the effect of gas flows on the roughness of the etched surface was studied and it was shown that the determining parameter of the angle deviation from the vertical is the gas flow and the crystallographic direction of these structures.

References

Ageev, O. A., Klimin, V. S., Solodovnik, M. S., Eskov, A. V., Krasnoborodko, S. Y. The study of influence of the gas flow rate to etched layer thickness, and roughness of the anisotropy field of gallium. arsenide is etched in the plasma chemical etching process. *Journal of Physics: Conference Series*, 2016, T. 741, № 1. p. 012178.

Klimin, V. S., M. S. Solodovnik, V. A. Smirnov, A. V. Eskov, R. V. Tominov, O. A. Ageev et al., *Proceedings of SPIE - The International Society for Optical Engineering,* 102241Z (2016).

Klimin, V. S., M. S. Solodovnik, A. V. Eskov, S. Y. Krasnoborodko et al., *Journal of Physics: Conference Series,* 741, 012178 (2016).

In: Applied Aspects of Nano-Physics …
Editors: Kirill Levine and Andrey Syrkov
ISBN: 978-1-53614-708-7

Chapter 15

NANOSTRUCTURES WITH QUANTUM WELL INGAAS/GAAS: EXPERIMENTAL AND THEORETICAL ANALYSIS OF TUNNEL CONTRIBUTION TO EMISSION OF CHARGE CARRIERS

Y. V. Ivanova** and V. I. Zubkov
SpbSETU "LETI", Saint-Petersburg, Russian Federation

ABSTRACT

The study of heterostructures with a single quantum well (SQW) $In_xGa_{1-x}As/GaAs$ was carried out by admittance methods, demonstrating two competing mechanisms of emission for carriers: thermionic and tunnel.

The dependence of thermionic peaks of conductance on the reverse bias has resonance character. On the other hand, we noticed the temperature independent plateau on conductance-temperature spectra, which is always related to the tunnel nature. We guess the observed effect is the resonant tunneling through the two-barrier potential, formed at the QW borders due to the doping.

Also, the broadening of conductance spectra with decreasing content x in SQW is detected. The results were analyzed using the numerical self-consistent simulation.

Keywords: single quantum well, admittance spectroscopy, tunneling, thermionic emission

---

* Corresponding Author's Email: yana@unix-server.su.

INTRODUCTION

Semiconductor heterostructures containing QW are widely used in different industries, particularly in the mineral and raw complex industries, as various light-emitting diodes, photodetectors and microwave devices. The main feature of such nanomaterials is the governing of their macro properties by quantum mechanical effects.

As we have shown [1], in these objects there are two competing mechanisms of charge carrier emission: thermionic emission and tunneling, which make different contributions to conductance at different temperatures.

EXPERIMENTAL SECTION

In this work, the isotype n-type heterojunctions with elastically strained single quantum well (SQW) $In_xGa_{1-x}As/GaAs$ (x = 0.065…0.29) grown by MOCVD were investigated. Thicknesses of the active layer of samples were 6.0…9.5nm.

The parameters and quality of the structures were controlled by HRXRD, local cathodoluminescence, etc. Conductance measurements were taken in temperature range from 10 to 375 K, voltage range of ± 40V, and frequency range of test signal 20Hz – 2MHz.

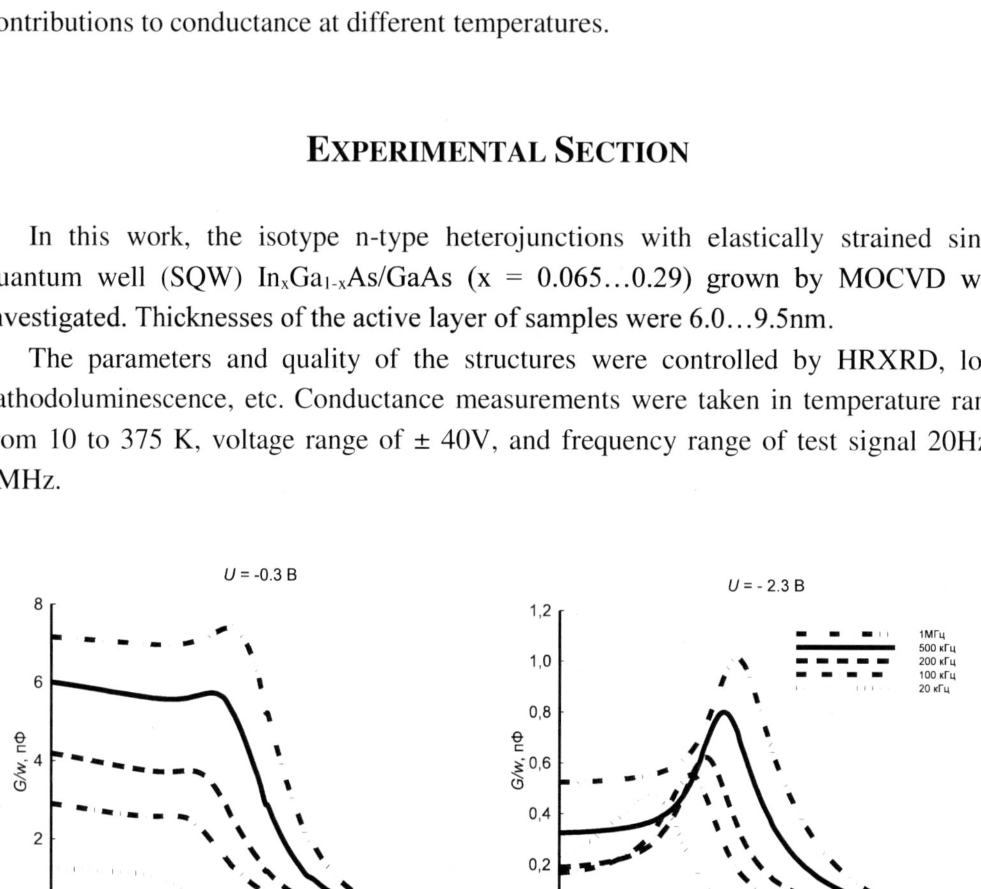

Figure 1. Admittance spectra of heterostructure with QW $In_{0.29}Ga_{0.71}As/GaAs$, obtained at U = -0.3V (no electric field in the QW) and at U = -2.3V (electric field approached the QW).

RESULTS AND DISCUSSION

The response in conductance spectra, corresponding to QW recharge, was registered only for QW with high In content in solid solution (x > 0.25). The temperature range for observation of response was 10 - 120K. Results for structure with x = 0.29 (the border of pseudomorphous growth) are presented in figure 1. The temperature-conductance (G-T) spectra are experienced strong modification in dependence on the reverse bias. Peaks are expressed at reverse bias -2.3V, which corresponds to the part of CV characteristics responsible for the beginning of intensive thermionic emission of charge carriers from quantization levels with the electric field penetrating into the QW. The observed phenomenon has resonance character and exists only in narrow voltage range – about 0.5V. In G-T experiments the "shelf" are observed at temperatures of 20 - 50K that demonstrates the existence of tunnel current, which got the maximal value at - 0.3V, and then had sharp fall in more than 10 times of value.

CONCLUSION

Numerical modeling by the self-consistent calculation of Schrödinger and Poisson equations shown that the top part of doped heterostructures with the QW on the band diagram represents a two-barrier structure in the energy region of quasi-resonances (E > 0) [1] with the opportunity for the resonant tunneling. Identical barriers have the maximum probability of resonance tunneling. At small reverse biases, the external electric field does not reach a space charge region yet, and the QW remains with symmetric barriers, so "shelves" on G-T spectra get the maximal values. On increasing the reverse bias the left-hand edge of the space charge region begins "to rise", a symmetry of barriers is broken and the tunnel transfer probability sharply falls. In the experiments, it is expressed by a sharp falling of the "shelf" at -1.3 V.

REFERENCE

[1] Zubkov, V. I. 2007. "Characterization of $In_xGa_{1-x}As/GaAs$ quantum-well heterostructures by C–V measurements: band offsets, quantum-confinement levels, and wave functions". *Semiconductors*, Vol. 41, N 3: 320 - 326.

In: Applied Aspects of Nano-Physics …
Editors: Kirill Levine and Andrey Syrkov
ISBN: 978-1-53614-708-7

Chapter 16

PROPERTIES OF ER–DOPED $(Na,K)NbO_3$ NANOFIBERS

Nadezhda P. Markova* and Alexander M. Grishin
Petrozavodsk State University, Petrozavodsk, Russia

ABSTRACT

In this chapter, we report static electrical and optical properties of Erbium doped $(Na,K)NbO_3$ nanofibers. 100 μm long and 100-200 nm in diameter nanofibers were sintered by the sol-gel calcination assisted electrospinning technique. Er doping with the concentration of 2 at.% provides readily detectable room-temperature broad-band photoluminescence centered at λ_{PL} = 0.55 and 0.98 □m being pumped, respectively, with 532 and 785nm lasers.

Keywords: electrospinning, ferroelectric fibers, electric properties, photoluminescence

INTRODUCTION

Ferroelectric sodium potassium niobate [$(Na,K)NbO_3$, hereinafter NKN] ceramics were patented and FDA approved (U.S. Food and Drug Administration) as a biocompatible material for implants [1]. Thorough toxicology tests showed there were no bacterial products (endotoxin) that would appear. Additionally, human monocytes were not negatively affected by the presence of NKN ceramics. Erbium doping promises new photonic applications to $(Na,K)NbO_3$ ceramics which is already abundantly supplied with

* Corresponding Author's Email: khomlyk@mail.ru.

piezoelectric, electrostrictive, low microwave loss voltage-tunable dielectric, and electro-optic properties.

Experimental Section

Fabrication of $Er:Na_xK_{1-x}NbO_3$ fibers was performed by the electrospinning method while formation of nanowires occurred in electrostatic field from the jet of polymer-precursor solution.

To prepare NKN precursor solution by the sol-gel method, sodium $NaO_2C_2H_3$ and potassium $KO_2C_2H_3 \times 3H_2O$ acetates were mixed in 2-methoxyethanol at room temperature and stirred for 1 hour to achieve a clear and transparent liquid mixture. Niobium ethoxide $Nb(OCH_2CH_3)_5$ and erbium nitrate pentahydrate $Er(NO_3)_3 \times 5H_2O$ were dissolved at room temperature in acetyl acetone, stirred for 1 hour, then added to the Na/K precursor solution, and continued stirring for 24 hours in a closed cap glass ware. Polyvinylpyrrolidone (PVP, 0.05g/ml) was added to the precursor mixture to prepare the solution for electrospinning.

Viscous polymer-precursor jet was ejected from a syringe pump NE-300 that feeds the PVP/NKN solution at a constant rate of 0.5ml/hour an electric field 1.8kV/cm between metallic needle and aluminum foil collector located at 15cm below the tip. The diameter of ejected fibers ranges from 300 to 350nm.

After the process of electrospinning, to remove a polymer binder, collected jelly-like fibers were dried at 100°C in nitrogen atmosphere and annealed in the high-temperature furnace OTF-1200X in air at 800°C. The rate of heating/cooling was 5°C/min [2]. The diameter of fibers after annealing varies between 100-200nm.

Results and Discussion

XRD pattern of Er:NKN fibers recorded with the Siemens D-5000 diffractometer validates the predominance of perovskite $Na_{0.35}K_{0.65}NbO_3$ – like phase.

Electric properties of Er:NKN nanofibers were investigated using measurement of current-voltage *I-V* characteristics. They were traced for the planar Au/Er:NKN/Au cell with 350μm thick Er:NKN fiber fabric placed onto glass-ceramic *Sitall* substrate as well as for the vertical sandwich Pt/Er:NKN(260μm)/Au structure. Circular 0.95 mm in diameter ohmic Au electrodes were thermally evaporated through a shadow mask on the top of nanofiber fabric specimens.

Two effects clearly demonstrate ferroelectric properties of Er:NKN fibers. The first is a clockwise hysterethis *I-V* loop at weak electric field in the planar Au/Er:NKN/Au cell.

The second was exhibited in the vertical sandvich Pt/Er:NKN/Au cell. A positive bias voltage shift occurs to nullify current at the descending branch of *I-V* curve. Also, Pt/Er:NKN/Au sandwich structure posseses a strong rectification of direct current as well as a resistance switching from low-resistance to high-resistance state with the resistance ratio of 10^2. Quasi-static *I-V* characteristics of Pt/Er:NKN/ Au diode correlate with the results obtained by impedance spectroscopy carried out at frequencies up to 1 MHz.

Intensive room temperature luminescence was observed by pumping fibers with 532 and 785nm lasers light. Radiation of 532nm produces intensive luminescence at λ_{PL} = 0.55μm and a weaker one within the λ_{PL} band from 0.96 to 1.11μm. Also, it excites Raman-active internal vibrations of NbO_6 octahedra. Radiation from 785nm laser results in the appearance of strong fluorescence lines with the wavelengths from 0.96 to 1.11μm.

Photoexcited luminescence adds to NKN fibers great potential to serve as multimodal bioimaging, tensile and torsion sensors, energy harvesting biocompatible nanogenerators, material for bioelectronic implants, electrically polarizable scaffolds for bioengineering, repair, and regeneration of damaged tissue.

This work was supported by the RF Ministry of Education and Science, project no. 16.5857.2017/8.9 (state program) and by the Flagship University Development Program of Petrozavodsk State University (2017–2021).

References

[1] K. Nilsson, J. Lidman, K. Ljungstrom, C. Kjellman, U.S. patent 6,526,984 (4 March 2003).

[2] A. M. Grishin, N. P. Markova, *J. Am. Ceram. Soc.*, Vol. 100 (2017), p. 1051.

In: Applied Aspects of Nano-Physics ...
Editors: Kirill Levine and Andrey Syrkov ISBN: 978-1-53614-708-7

Chapter 17

INVESTIGATING THE INFLUENCE OF TECHNOLOGICAL MANUFACTURING CONDITIONS ON STRUCTURAL CHARACTERISTICS OF AMORPHOUS SILICON LAYERS OF HETEROJUNCTION SOLAR CELLS

Aleksei D. Maslov**, Ekaterina V. Bezuglaya and Nikolay V. Vishnyakov
Ryazan State Radio Engineering University,
Ryazan, Russia

ABSTRACT

The heterojunction silicon solar cells based on a-Si/c-Si were investigated via the Raman spectrocopy. We analyzed the Raman spectra to determine structural characteristics of amorophous films that allow us to make a conclusion about the degree of disordering of amoprhpous silicon layers on the basis of estimating the angle of a coupling deviation. It is shown that the degree of disordering strongly depends on technological conditions of manufacturing an amorphous layer. Correlations between such technological parameters as substrate temperature, deposition time, plasma treatment of the surface and structural properties of a-Si are found.

Keywords: heterojunction silicon solar cells, Raman spectroscopy, structural characteristics, technological conditions, amorphous layer

---

* Corresponding Author's Email: maslov.a.d@mail.ru.

INTRODUCTION

Heterojunction solar cells based on a-Si/c-Si junction are promising structures for effective cheap terrestrial use. However disorder amorphous layers might restrict efficiency of such structures due to the recombination by dangling bonds, parasitic absorbance and resistivity of these layers. Therefore it is necessary to control such parameters of a-Si layers as thickness, doping concentration, density of recombination states etc. Since density of states strongly depends on structural features of amorphous web it is necessary to investigate and control structural characteristics. These parameters depend on technological conditions of amorphous layer manufacturing. Thus the aim of this paper is determination of correlations between technological conditions and structural characteristics of a-Si.

For investigation of structural characteristics we used widespread Raman spectroscopy method that is often used to study structural information [1]. Raman spectrum of amorphous silicon is a complicated spectrum that consist of four peaks corresponding TA - transverse acoustical, TO - transverse optical, LO - longitudinal acoustic and LO - longitudinal optical vibrations of non-crystalline web. Typical spectrum of amorphous silicon is shown in Figure 1 [2].

Compared to other methods of investigating structural properties, Raman spectroscopy is a more sensitive method to small fluctuations in structure within the first coordination sphere [3].

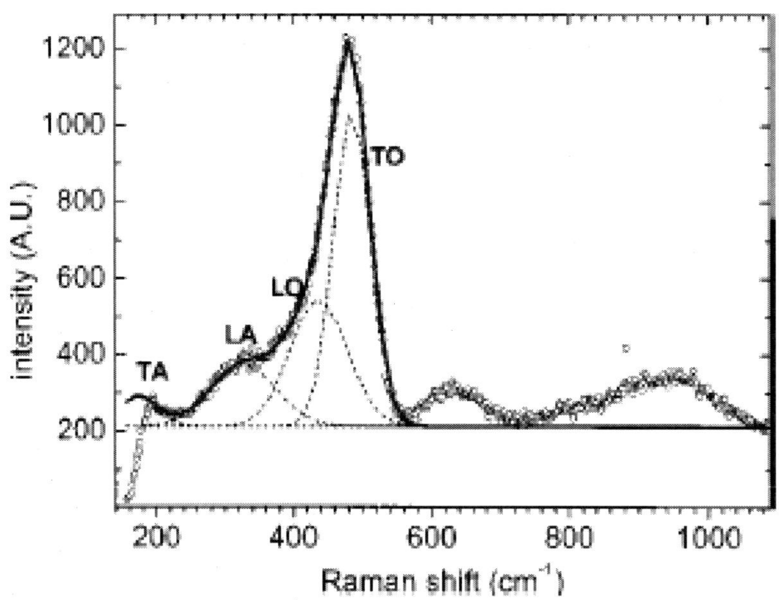

Figure 1. Typical spectrum of amorphous silicon [2].

For example, measuring width of TO-peak allows to calculate root-mean-square deviation of the coupling angle from the tetrahedral position of the silicon crystal using Beeman's linear empirical relation [3]:

$$\Gamma = 15+6\Delta\theta \quad (1)$$

where Γ- width of the peak, cm^{-1}; Δθ - root-mean-square deviation, degrees.

EXPERIMENTAL RESULTS

We investigated four testing a-Si:H/ c-Si solar cells manufactured with different technological conditions described in Table 1.

Raman spectra are obtained using automated AFM-Raman, SNOM and TERS system Ntegra Spectra (NT-MDT, Russia). We used a 532nm solid state laser with power 0.4mW.

Table 1. Technological conditions of amorphous silicon

№	Substrate temperature, °C	Deposition time of a-Si:H, hours	Thickness. um	Notes
1	220	5	3	
2	170	5	3	
3	230	2,5	1,5	
4	230	2,5	1,5	Plasma treatment of surface

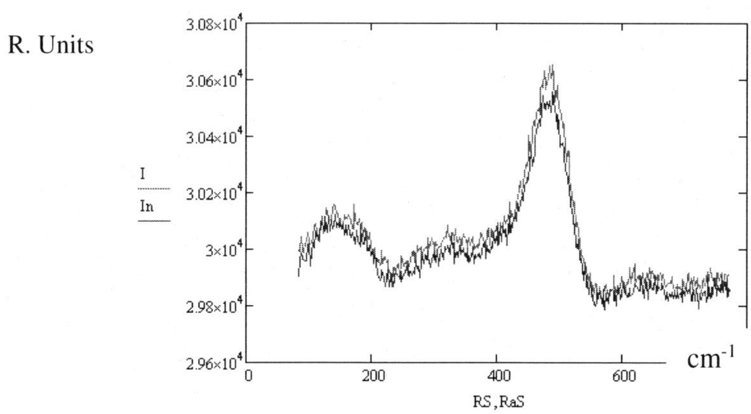

Figure 2. Raman spectra of samples №1 and №2. ___ Sample № 1. ___ Sample № 2.

The obtained Raman spectra are shown in Figures 2 and 3.

In addition to Beeman's empirical relation (1), the degree of ordering might be determined by estimation of TO-peak position and the ratio of TA/TO peak intensities [3].

Results of calculating the ordering degree of amorphous layer are shown in Table 2.

Disordering of amorphous web should influence on increasing of deviation of the coupling angle from the tetrahedral position appropriate to crystalline silicon, the intensity ratio of the TA/TO peaks, and the shift of the TO peak position.

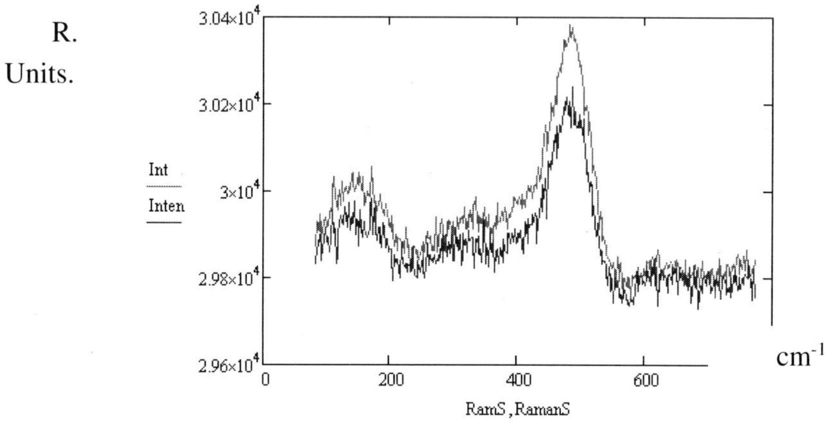

Figure 3. Raman spectra of samples №3 and №4. ___ Sample № 3. ___ Sample № 4.

Table 2. Results of calculating structural parameters

Parameter Sample	Angle of coupling deviation, degrees	TO-peak position, cm^{-1}	Ratio of TA/TO peak intensities
1	8,6	486	0,983
2	8,8	486	0,986
3	10,6	486	0,988
4	10,8	474	0,991

Thus changes in structural characteristics of the samples are correct [4]. The samples №1 and №2, manufactured with different substrate temperature have different angle of coupling deviation and TO/TA peak ratio: sample №1 has lower values because this film is less disordered due to the lower substrate temperature (Table 1).

Increasing of angle of coupling deviation and TO/TA peak ratio of samples №3,4, obtained with higher substrate temperature than samples №1,2, is related with influence of deposition time on formation of amorphous web.

In the case of comparison of samples №3 и №4, manufactured with substrate temperature 230°C, sample №4 with surface bombarded by Ar^+, according to Table 2 is more disorder film due to the increase of defects states. It is confirmed by shift of TO-peak.

CONCLUSION

We found correlations between technological conditions of amorphous silicon layer and structural characteristics using Raman spectroscopy method. Substrate temperature, deposition time and plasma treatment of surface are considered. The degree of disordering of amorphous web is estimated by Beeman's empirical relation, TO-peak position and the ratio of TA/TO peak.

The experiments are made using equipment of Regional center of probe microscopy of Ryazan state radio engineering university.

REFERENCES

[1] John R. Ferraro, Kazuo Nakamoto and Chris W. Brown. *Introductory Raman spectroscopy*. Elsevier, 2003. p. 434.

[2] C. Smit, R. A. C. M. M. van Swaaij, H. Donker, A. M. H. N. Petit, W. M. M. Kessels, M. C. M. van de Sanden. Determining the material structure of microcrystalline silicon from Raman spectra. *J. Appl. Phys.,* 2003. Volume. 94.pp. 3582 - 3588.

[3] R. L. C. Vink, G. T. Barkema, W. F. van der Weg. *Relation between Raman spectra and Structure of Amorphous Silicon*. 2000. Vol. 19. pp. 156 - 162.

[4] Brodsky, Marc H. *Amorphous semiconductors*. Springer-Verlag, 1979. p. 337.

In: Applied Aspects of Nano-Physics …
Editors: Kirill Levine and Andrey Syrkov
ISBN: 978-1-53614-708-7

Chapter 18

The Influence of Technology on the Properties of Thermoelectric Materials Produced by Directional Crystallization

Sergey A. Nemov[1,2], Arseny A. Rulimov[1,*], Alexandr V. Shchegol'kov[3] and Anton E. Demchenko[1]

[1]Peter the Great Saint Petersburg Polytechnic University, Saint Petersburg, Russia

[2]Saint Petersburg Electrotechnical Univeristy, Saint Petersburg, Russia

[3]Tambov State Technical University, Tambov, Russia

Abstract

Attempts to improve the properties of $Bi_xSb_{2-x}Te_3$ solid solution produced by directional crystallization are still going on. It is argued that the addition of selenium increases the thermoelectric properties of $Bi_xSb_{2-x}Te_3$ samples by reducing the number of defects in Sb_2Te_3 crystal lattice. Besides we confirm the heterogeneity of the thermoelectric coefficient in $Bi_xSb_{2-x}Te_3$ samples' cross section.

Keywords: $Bi_xSb_{2-x}Te_3$, Vertical zone melting, Heterogeneity of the thermoelectric properties, Selenium

* Corresponding Author's Email: rulimmmov@gmail.com

INTRODUCTION

Despite the long-standing reputation of low-temperature thermoelectric materials based on Sb and Bi tellurides, nowadays the theoretical base of the enterprises is being actively updated with the information of such a solid solution like $Bi_xSb_{2-x}Te_3$. The main problem in the production of corresponding materials is the high sensitivity to technological parameters, which unjustified variation entails loss of both thermoelectric and mechanical properties.

MATERIALS AND METHODS

$Bi_xSb_{2-x}Te_3$ solid solution is widely used in thermal converters with a working temperature range from 300K to 600K, wherein the contain of antimony telluride varies from 70 to 80mol. % of the total amount of tellurides. Directional crystallization methods have established themselves as basic for the production of thermoelectric materials including the above-mentioned composition. Growth rate anisotropy makes it possible to grow crystals with a directed structure, in which the value of the thermoelectric efficiency ZT is maximized along the growth direction [1].

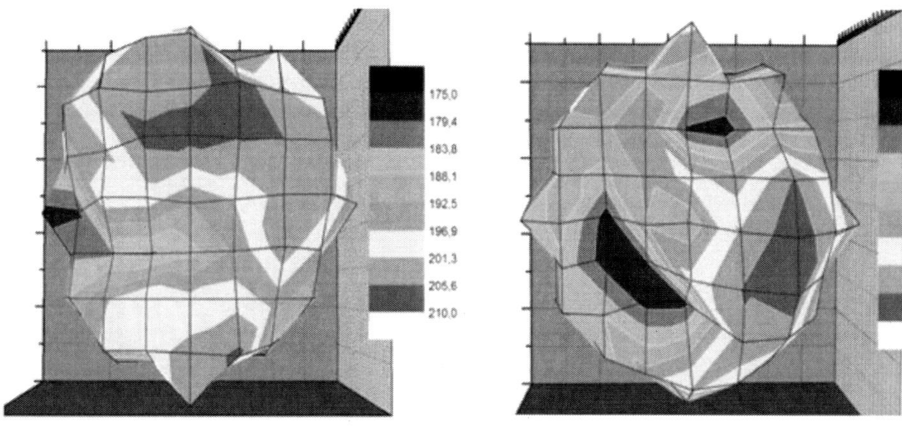

Figure 1. The distributions of the thermoelectric coefficient [μV/K] in two $Bi_xSb_{2-x}Te_3$ samples' cross section.

DISCUSSION

During the numerous experiments we have investigated $Bi_xSb_{2-x}Te_3$ samples with the hole conductivity to detect the possible heterogeneity of thermoelectric properties. The samples had been produced by vertical zone melting. Particularly we have attempted to plot a distribution of the thermoelectric coefficient in the samples' cross section using thermal measuring probe. Typical distributions are shown in Figure 1.

For precise adjustment of the composition and, as a consequence, useful properties of the samples and final products, it is expedient to introduce selenium into the composition, i.e., create quaternary (Bi_xSb_{2-x}) ($Te_{3-y}Se_y$) solid solution. The fact is that selenium additive smoothly reduces the hole concentration and the number of defects in Sb_2Te_3 crystal lattice, which causes an increase in the hole mobility. All of the above has a positive effect on the values of σ and α [2, 3].

RESULTS AND CONCLUSION

During the research, we have confirmed and described the heterogeneity of thermoelectric properties in $Bi_xSb_{2-x}Te_3$ samples produced by vertical zone melting. Consistent patterns in the distribution of the thermoelectric coefficient along the samples' cross section were not found. This can be explained by the presence of compositional fluctuations at the crystallization front.

Small selenium additives to the charge improve the thermoelectric properties of $Bi_xSb_{2-x}Te_3$ samples.

REFERENCES

[1] Lavrent'ev, M. G., Osvensky, V. B., Pivovarov, G. I., Sorokin, A. I., Karataev, V. V., Bublik, V. T., Tabachkova, N. Yu. (2015). Mechanical properties of bismuth and antimony solid solutions produced by directional crystallization and powder metallurgy. Thermoelectric materials and their applications. *Reports of the Interstate Conference*. The Ioffe Institute (St. Petersburg).

[2] Ivanova, L. D., Sidorov, Yu. A. (1998). Electrical properties of Se-doped $Bi_{0.5}Sb_{1.5}Te_3$ crystals. *Inorganic Materials,* 34 (3): 222-224.

[3] Luk'yanova, L. N., Kutasov, V. A., Popov, V. V., Konstantinov, P. P. (2004). Galvanomagnetic and thermoelectric properties of p-$Bi_{2-x}Sb_xTe_{3-y}Se_y$ solid solutions at low temperatures (< 220 K). *Physics of the Solid State,* 46 (8): 1404-1409.

In: Applied Aspects of Nano-Physics …
Editors: Kirill Levine and Andrey Syrkov
ISBN: 978-1-53614-708-7

Chapter 19

EXPERIMENTAL STUDY OF CARRIER DISTRIBUTION IN CVD SINGLE-CRYSTAL DIAMOND WITH DELTA-DOPED LAYERS

L. M. Shestakova **[*]** ***and V. I. Zubkov***
SpbSETU “LETI”, Saint-Petersburg, Russian Federation

ABSTRACT

Properties of multi-delta-doped diamond are studied. The goal of this research is to measure the charge carriers' concentration profile in single crystal diamonds with four delta-doped layers. Addmitance spectroscopy was used as the principal method to measure the properties of samples at different temperatures and applied biases. The measurements showed that only the first (shallow) δ-layer was confidently observed, especially at elevated temperatures (350-450K). Besides, the concentration profiles included a weak response from the second delta layer, significally shifted from its true location. A general conclusion is made concerning the observed charge carriers' distribution.

Further study is needed to determine the responses from the deeper delta-doped layers.

Keywords: single-crystal diamond, admittance spectroscopy, delta-doping

[*] Corresponding Author's Email: l94shes@yandex.ru.

INTRODUCTION

The search for more advanced materials and research of the methods for their application is an actual task in microwave electronics, power electronics, radiophotonics, etc. The most promising material for such applications is semiconducting diamond. However, due to the problem of charge carriers scattering on ionized impurities, the delta-doping becomes the main method for obtaining the super-high frequencies devices based on diamond.

EXPERIMENT SECTION

The paper presents the results of study a single-crystal diamond sample with four delta-doped layers grown by CVD method. The period of delta-doped layers is about 60nm.

Measurements of current-voltage and capacitance-voltage characteristics were obtained using a computer-controlled set up of admittance spectroscopy.

RESULTS AND DISCUSSION

For measurements below 230K, the sample behaves like a dielectric. Due to the incomplete impurity ionization in diamond, the charge under the rectifiyng barrier is collected from different depths at different temperatures; therefore at 300K the depletion region overlapped the delta-doped layers at zero bias. As the temperature rises, the depletion region narrows, so that at $T > 400$K it is possible in CV-measurements to register the distribution of charge carriers in the delta-layer region, Figure 1.

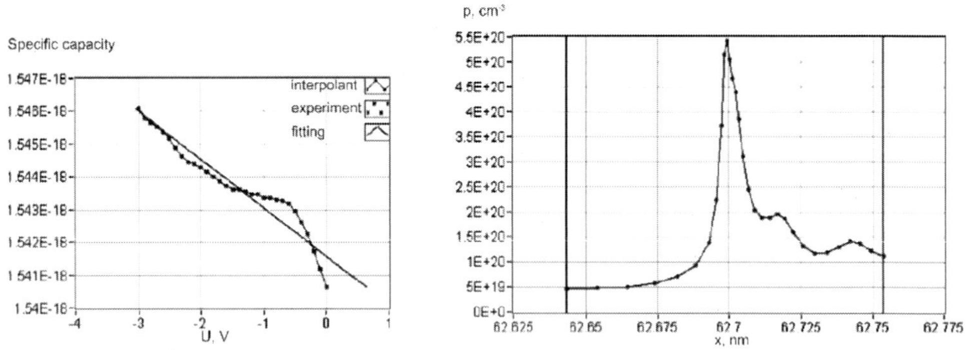

Figure 1. CV characteristics and concentration profiles in sample with delta-layers (450 K).

The position of the concentration peak (62.7nm) corresponds fairly well to the technologically specified position, taking into account the hardware error of the CV-method. The peak amplitude reaches $5.5 \cdot 10^{20} cm^{-3}$.

As it can be seen, the profiles also include the responses from the deeper delta-doped layers, but their shape and amplitude need further investigation.

CONCLUSION

The measurements of concentration profiles in the region of delta-doped layers in single-crystal diamond were carried out for the first time in the practice of capacitance measurements. The object of further research will be the simulation of concentration profiles by the method of self-consistent solution of the Schrödinger and Poisson equations.

REFERENCES

[1] Zubkov, V. I.; Solomnikova, A. V.; Post, J. E. et al. *Diam. Relat. Mater.*, **2017.** *72.* 87 - 93.

[2] Zubkov, V. I.; Kucherova, O. V.; Bogdanov, S. A. et al. *J. of Appl. Phys.*, **2015**. *118.* 145703.

In: Applied Aspects of Nano-Physics ...
Editors: Kirill Levine and Andrey Syrkov

ISBN: 978-1-53614-708-7

Chapter 20

POROUS SILICON-BASED NANOSTRUCTURES FOR SOLAR CONCENTRATOR SYSTEMS

R. S. Smerdov[1,*], A. S. Mustafaev[1], Yu. M. Spivak[2] and Vyacheslav A. Moshnikov[2]

[1]Saint Petersburg Mining University, Saint Petersburg, Russia
[2]Saint-Petersburg State Electrotechnical University, Saint Petersburg, Russia

ABSTRACT

The investigation of porous silicon/silver nanostructured composite layers (por-Si/Ag) via Raman spectroscopy technique has been carried out. Interpretation of por-Si/Ag Raman spectra has been proposed. Data from layers acquired by varying anodization time showed a significant shift of a Raman peak primarily located at the frequency expected for bulk silicon towards lower energy. In addition to the single crystalline peak a number of low-frequency bands emerged in the Raman spectra of silver-functionalised porous silicon. These effects have been interpreted as due to the influence of functionalizing silver nanoparticles and synthesis parameters on the structure and properties of composite material.

Keywords: solar concentration systems, PETE, Raman spectroscopy, porous silicon, silver nanoparticles, surface plasmon resonance, anodization

* Corresponding Author's Email: rostofan@gmail.com.

INTRODUCTION

Porous silicon (PS) and PS-based composite nanostructures are of great interest due to unique properties of such materials including photoluminescence, electroluminescence and the dependence of refraction index on porosity. It is well established that porous silicon is a promising material for numerous applications associated with micro-, nano- and optoelectronics, medicine (biosensors, targeted drug delivery), gas sensing and so on [1] by virtue of its high specific surface area and extensive capability for surface functionalization with various structures.

The use of semiconductor (GaN) cathodes in photon-enhanced thermionic emission systems has proved to be effective although only a few (less than 1%) of the incident photons exceed the 3.3 eV GaN band gap [2]. Due to the aforementioned drawback further research in the field of semiconductor materials and nanocomposites is required. Porous silicon (PS) and PS-based nanostructures in particular exhibit nearly optimal parameters in terms of the band gap (1.1 eV).

The band gap modification for PS-based composite nanostructures is possible in the wide range of 1 eV and 3 eV due to the existence of quantum confinement phenomenon and the extensive capabilities of PS surface functionalization thus providing us with an applicable material for solar concentration and thermionic energy conversion systems cathode fabrication.

Composite PS – metal materials can also act as a promising substrate for identification and characterization of various biological and chemical structures due to the surface enhanced Raman scattering effect (SERS) caused by excitation of localized surface plasmons in metal nanoparticles.

EXPERIMENTAL SECTION

PS layers were obtained by anodization of monocrystalline (111) n-type silicon wafers with resistivity of 4.5 $\Omega\cdot$cm. Anodization time varied in the range of 0.5 to 40min. The deposition of silver was carried out from an aqueous solution of silver nitrate in a three-electrode electrochemical cell. The silver chloride electrode was used as the reference electrode. Porous silicon acted as a working electrode.

The Raman spectrum of initial porous silicon matrix is characterized by a peak located at 521cm^{-1} that is indicative of the initial silicon sample crystalline structure.

According to Cardona [3] the size of a nanocrystalline Si cluster corresponding to a certain Raman shift could be estimated by using equation

$$d = \frac{2\pi}{10}\left(\frac{B}{\Delta\omega}\right)^{1/2}, \tag{1}$$

where B = 224cm^{-1}, Δω is the frequency down shift of the sample referring to the c-Si line (521cm^{-1}). Calculated sizes of Si nanocrystals turn out to be 8nm, 8nm, 6.2nm, 6.2nm, 5.2nm with respect to 0.5 min, 1 min, 10 min, 20 min, and 40 min anodization times.

The Raman bands at 95cm^{-1} and 146cm-^{1} have been attributed to the silver lattice vibrations. Duval et al. [4] states that Raman spectra of both metal nanoparticles and composite materials based on those contain a low-frequency peak the nature of which is related to the mechanical oscillations of nanoparticles under the influence of incident optical radiation. Herein by using Lamb theory for nanoparticle vibrational modes it is possible to derive an equation for the frequency of the lowest-energy spherical mode as

$$\omega_2 = 0.85\frac{v_t}{dc}, \tag{2}$$

where v_t = 1600m/s, is the transverse sound velocity, a value averaged on the different directions in the crystal [4], d - is the diameter of a particle, c - is the speed of light in vacuum, ω is the observed peak position that leads to a diameter of Ag nanoparticles deposited on PS surface of approximately 0.7nm.

CONCLUSION

The porous silicon matrix structure before the deposition of functionalizing Ag nanoparticles closely resembles monocrystalline silicon structure. The deposition of silver nanoparticles results in the deformation of silicon lattice, accompanied by the appearance of the Raman bands being uncharacteristic for the unmodified porous silicon. Functionalization of PS matrix corresponds to the emergence of silicon nanocrystallytes, resulting in a shift of the peak intrinsic to monocrystalline silicon (521cm^{-1}) towards lower energy.

The estimation of the Si nanocrystallytes scale is performed according to Cardona theory [3] with anodization time of a silicon matrix as a parameter.

The origin of low-energy Raman bands has been related to the formation of Ag nanoparticles [4] (from 0.7nm in diameter) on the surface of a PS matrix. The increase in anodization time of silicon samples gives rise to further increase in the thickness of the porous layer and the pore diameter.

The authors are grateful to V.S. Levitsky (R&D Center TFTE, Saint Petersburg, Russia) for the Raman spectra measurements of nanostructured PS/Ag layers.

REFERENCES

[1] Smerdov, R. S., Loboda, V. V., Spivak, Yu. M., Moshnikov, V. A. 2016 St. Petersburg State Polytechnical University Journal Computer Science. *Telecommunications and Control Systems*, 3(247) 12.

[2] Schwede, J. W., Bargatin, I., Riley, D. C., Hardin, B. E., Rosenthal, S. J. Photon-enhanced thermionic emission for solar concentrator systems. *Nature Materials*, 2010. N 9, p.762 - 767.

[3] Cardona, M. 1989 *Superlattices and Microstructures*, 5 27.

[4] Duval, E., Boukenter, A. and Champagnon, B. 1986 *Phys. Rev. Lett.*, 56 2052.

In: Applied Aspects of Nano-Physics …
Editors: Kirill Levine and Andrey Syrkov
ISBN: 978-1-53614-708-7

Chapter 21

ATOMIC FORCE MICROSCOPY OF NANOCOMPOSITES BASED ON ZINC OXIDE WITH DIFFERENT ADDITIVES

Evgeniya V. Maraeva[1,*], Vyacheslav A. Moshnikov[1], Nadezhda D. Yakusheva[2], Igor A. Pronin[2] and Igor A. Averin[2]
[1]Saint Petersburg Electrotechnical University "LETI", Saint-Petersburg, Russian Federation
[2]Penza State University, Penza, Russian Federation

ABSTRACT

This chapter is concerned with the investigation of nanomaterials based on ZnO with different additives (Fe, Cu).

A correlation between technological modes of production (annealing temperature, composition) and the values of the fractal dimension of the samples is discussed.

Keywords: porous nanoparticles, atomic force microscopy, fractal dimension, zinc oxide

* Corresponding Author's Email: jenvmar@mail.ru.

INTRODUCTION

Currently, the nanocomposites based on metal oxides are widely used for the creation of gas sensors with a percolation structure [1-3]. The devices based on such structures allow one to reach extremely high values of gas-sensitive instances [4]. One of the most important ways to study such structures is fractal analysis based on data with the use of atomic force microscopy (AFM).

There are different approaches to estimate the fractal dimension of the nanostructures, such as the method of cube counting, triangulation method, power spectrum method etc. However, all of these methods depend on many factors (for example, the size of the scan area). All of them can be implemented in the graphic program «Gwyddion» for scanning probe microscopy (SPM) data analysis.

EXPERIMENTAL SECTION

The main purpose of this work was to study the peculiarities of the surface relief in the nanocomposites based on zinc oxide and to determine the influence of annealing temperature and composition of investigated materials on the value of their fractal dimension. The stated purpose implies the following objectives:

1. To study the nanocomposites based on zinc oxide with different additives (Fe, Cu) with the use of scanning probe microscopy;
2. To identify the relationship between technological modes of production (composition, annealing temperature) and the values of fractal dimension of the samples.

The study of surface morphology of nanocomposites based on zinc oxide were carried out by means of scanning probe microscopy with atomic - force microscope NTEGRA (NT-MDT, Russia) with the use of silicon probes NSG with a radius of 10 nm in tapping mode. Evaluation of the fractal dimension was carried out in a graphics program for SPM data analysis «Gwyddion». The methods of cube counting and triangulation method based on the analysis of AFM images were used.

RESULTS AND DISCUSSION

For research purposes, a set of samples based on zinc oxide with additives of Fe and Cu, heated at different temperatures (420K, 870K) was obtained [4]. With the use of atomic force microscopy method it was found that all the samples were characterized by the presence of close-packed grain system.

Table 1. The results of fractal analyses

Sample number	1a	2a	3a	1b	2b	3b
The composition of the sample	ZnO	ZnO (Fe)	ZnO (Cu)	ZnO	ZnO (Fe)	ZnO (Cu)
Annealing temperature, K	870	870	870	420	420	420
Fractal dimension (triangulation method)	2,54	2,60	2,55	2,58	2,58	2,33
Fractal dimension (method of cube counting)	2,47	2,52	2,46	2,47	2,48	2,36

From the morphological point of view the layers without additives of Cu and Fe were uniform despite of the processing conditions, in all parts of the surface the same relief was observed. The range of heights at different areas of a surface was 800 – 1200nm.

In some cases (the samples with Cu and Fe additives) the surface contained inclusions that represented the core with radiating channels. The surface areas free from impurities were characterized by typical samples relief. For all the patterns the calculation of fractal dimension values was carried out with the use of the program “Gwyddion”. The results are shown in Table 1. It can be seen from the Table 1 that for ZnO (Cu) pattern which was heat-treated at T = 420K the surface structure was characterized by the lowest value of the fractal dimension.

Conclusion

Thus, the atomic force microscopy data analysis showed that the presence of alloying elements (Fe, Cu) primarily affects the distribution of grains on the surface of the zinc oxide nanocomposites. The variation of the heat treatment regimes in the range from 420 K to 870K slightly affects the surface morphology of the samples. The lowest value of the fractal dimension characterizes the surface of ZnO (Cu) nanocomposites, annealed at T = 420K.

For other types of the samples, the fractal dimension varies slightly depending on the conditions of synthesis and depends mainly on the method of calculation.

ACKNOWLEDGMENTS

The reported study was supported by the Ministry of Education and Science of the Russian Federation within the framework of the project № 16.897.2017/ПЧ.

REFERENCES

[1] Relationship between the photocatalytic and photoluminescence properties of zinc oxide doped with copper and manganese / Pronin, I. A., Donkova, B. V., Dimitrov, D. Tz., Averin, I. A., Pencheva, J. A., Moshnikov, V. A.. *Semiconductors,* 2014. vol. 48, № 7, p. 842 - 847.

[2] Metal oxide SnO_2 - ZnO - SiO_2 films prepared by sol-gel / Abrashova, E. V., Kononova, I. E., Moshnikov, V. A.. *Smart Nanocomposites,* 2013. vol, 4, № 2, p. 1 - 7.

[3] Study of porous sol-gel nanocomposites based on silicon dioxide and tin dioxide modified by fullerenol $C_{60}(OH)_n$ (n = 22-24) / Maraeva, E. V., Istomina, M. S., Moshnikov, V. A., Nalimova, S. S., Maximov, A. I., Alexeev, N. I., Semenov, K. N.. *Journal of Physics: Conference Series,* 2016. vol. 690, p. 012031.

[4] Theoretical and experimental investigations of ethanol vapour sensitive properties of junctions composed from produced by sol-gel technology pure and Fe-modified nanostructured ZnO thin films / Pronin, I. A., Averin, I. A., Yakushova, N. D., Dimitrov, D. T., Krasteva, L. K., Papazova, K. I., Chanachev, A. S., Bojinova, A. S., Georgieva, A. T., Moshnikov, V. A.. *Sensors and Actuators A: Physical*, 2014. vol. 206, p. 88 - 96.

In: Applied Aspects of Nano-Physics …
Editors: Kirill Levine and Andrey Syrkov
ISBN: 978-1-53614-708-7

Chapter 22

NON-DESTRUCTIVE EVALUATION OF STRENGTH NANOFEATURES OF CONSTRUCTIONAL MATERIALS

***V. V. Nosov*[1,2]*, I. V. Matviian*[1,*] *and A. R. Yamilova*[1]**
[1]Saint-Petersburg Mining University, Saint Petersburg, Russia
[2]Peter The Great Saint Petersburg Polytechnic University, Saint Petersburg, Russia

ABSTRACT

Within a period of manufacturing and exploitation of different materials, complex physicochemical processes of interaction between their individual components can be observed. The revealing of such mechanisms helps to understand its distinct features and to optimize technological processes of manufacturing materials with certain characteristics. The solution could be based on the interpretation of results of acoustic emission (AE) tests from the perspective of a kinetic model of AE for heterogeneous materials.

Keywords: acoustic emission, heterogeneous materials, model of acoustic emission

[*] Corresponding Author's Email: matviian_ilia@mail.ru.

INTRODUCTION

The research technique lies in the experimental defining of the effect of different technological and exploitation factors on quantities of parameters concerning the AE model's coefficients.

RESULTS AND DISCUSSION

The presented model of timing dependence on the number of AE impulses has the following view:

$$N_{\sum}(t)=V\iiint_{\Delta t,f,u}\Phi(\Delta t,f,u)dudfd\Delta t\cdot C_0\int_{\mu}^{\mu+\Delta\omega}\psi(\omega)\left\{1-\exp\left[-\int_0^t dt'/\theta\left(U_0,\omega(t')\right)\right]\right\}d\omega \quad (1)$$

where

$\theta(U_o,\omega(t'))=\tau_o\exp\{[U_o-\gamma\sigma(t')]/(KT)\}$

Every parameter of the model [1] has its specific physical nature and depends on distinct factors that allow it to reveal mechanisms of impact for these factors on a material's features:

- parameter $V\iiint_{\Delta t,f,u}\Phi(\Delta t,f,u)dudfd\Delta t\cdot C_0=k_{AE}C_0$ characterizes amount of AE sources which are literally structural elements which we can "hear" via AE equipment during the process of destruction;
- correspondence of the variables of the $\Psi(\omega)$ function characterizes the degree of inhomogeneity of a material's mechanical state at a molecular level;
- parameter U_o (activation energy of the destruction process of molecular links) doesn't depend on the state of the material structure and is defined through characteristics of interatomic interaction (chemical ties) of a structural element;
- parameter γ (activation volume) is characteristic of a molecular nanostructure of material and it is a faintly sensitive parameter to its chemical nature.

The point of the survey lays in defining the systematical variability of the parameters within distinct technological changes and exploitation factors, and was based on the possibility of operational evaluation for these parameters due to the results of acoustic emission tests.

Different samples of heterogeneous materials were exposed to destructive AE tests in a regimen of steady loading with a constant rise in the speed of tension. It was noticed during the tests that the view of timing dependences N_{Σ} (t) were affected by such factors as size of filling material particles, time of isolation of the samples after their manufacturing, heat treatment and chemical saturation of the uppermost layers, which can be explained by the changing of degree in structural inhomogeneity of materials.

REFERENCES

[1] Nosov V.V. *Diagnostics of machines and equipment. Textbook.* Saint-Petersburg: Lan', 2016, p. 376 (in Russian).

[2] Nosov V.V., Matviyan I.V., *Mechanics of non-uniform materials. Textbook*. Saint-Petersburg: Lan', 2017, p. 276 (in Russian).

[3] Nosov V.V. *Mechanics of composite materials. Laboratory and practical exercises. Textbook.* Saint-Petersburg: Lan, 2013, p. 240. (in Russian).

[4] Nosov V.V. *Defektoskopiya* [*Defender*], 2014, № 12, 24-35. (in Russian).

[5] Nosov V.V., Potapov A.I. Acoustic Emission Testing of the Strength of Metal Structures under Complex Loading. *Russian Journal of Nondestructive Testing*, 2015, Vol. 51, No. 1, pp. 50–58.

[6] Nosov V.V. On the Principles of Optimizing the Technologies of Acoustic-Emission Strength Control of Industrial Objects. *Russian Journal of Nondestructive Testing*, 2016, 2016, Vol. 52, No. 7, pp. 386–399.

In: Applied Aspects of Nano-Physics ...
Editors: Kirill Levine and Andrey Syrkov
ISBN: 978-1-53614-708-7

Chapter 23

MODIFICATION OF ZINC FILMS BY CHEMICAL AND PHYSICAL METHODS

Vladimir V. Tomaev[1,*], Vladimir A. Polyshyk[2], Nikolai S. Pshchelko[3,†], Kirill Levine[1] and Yuri I. Kuzmin[1]

[1]Saint Petersburg Mining University, Saint Petersburg, Russia
[2]Admiral Makarov State University of Martitime and Inland Shipping, Saint-Petersburg, Russia
[3]Military Academy of Communications named after S. M. Budyonny, Russia

ABSTRACT

The method of thermal treatment of zinc films in non-pressurized volume was developed and evaluated. Zinc films on a glass substrate were subjected to thermal treatment in the presence of selenium vapors in the air, and combined thermal-electric field treatment. Morphology was assessed by SEM. Structural changes in initial and modified Zn films were discussed. Size and statistical parameters of zinc nanocrystals were determined. By the suggested method, it was possible to control zinc films morphology and composition in a wide range.

Keywords: zinc films, surface morphology, nanocrystals of zinc and zinc selenide, selenium vapors, new phase nucleation

* Corresponding Author's Email: tvaza@mail.ru.
† Author's Email: nikolsp@mail.ru.

INTRODUCTION

Zinc selenide (ZnSe) is known by its photo-resistive, photo- and electro-luminscent properties. Due to its high transparency in optical and infrared regions, it is used for manufacturing entry lenses and windows in optoelectronic devices [1].

The problem of obtaining, studying and usage of nanocrystalline films based on zinc halcogenides is on the edge of study [2-4].

In [5] the idea of optimization of photosensitive lead selenide (PbSe), by heating precursor films in the presence of selenium powder in the air in a non-hermetically closed vessel was realized. As a result of this procedure it was possible to obtain PbSe films with infrared sensitive capability at ambient temperatures (without cooling).

Polycrystalline ZnSe can be obtained by interaction of zinc with selenium vapors [5]. Monocrystals of cubic singonia were grown from vapor phase and melt, while monocrystals of hexagonal singonia – only from the vapor phase. ZnSe films were obtained by thermal evaporation of the compound and condensation on substrate heated to 150 - 250°C.

MATERIALS AND METHODS

Method in [5] describes chemical precipitation of PbSe from multicomponent mixtures under next following conditions:

- films are placed at non-hermetically close vessel, where certain ratio between surrounding volume and surface area of heating films is maintained, (usually ratio of total surface area to treating is 20-30.
- heating is carried out in the presence of metallic selenium.

Heating of Se in powder form allows saturating vessel with vapors of high partial pressure. Diffusion at temperatures of treatment creates conditions of controlled Se penetrating to lead. Varying experimental conditions allows varying Se concentration, decreases percentage of overoxidated sites and simultaneously decrease number of Se vacancies in PbSe crystalline lattice.

Experimental Section

Depositing Selenium on Zinc

In this chapter, the same method as described above was applied to treat zinc films by Se vapors.

Se surface-modified Zn films were obtained as a result of chemical interaction of Zn surface with Se vapors.

Izobaric conditions from one side, isolating from ambient atmosphere from the other, effectively decrease amount of oxygen during the reaction with simultaneous increasing of Se partial pressure. Oxygen becomes effectively consumed at initial stages of treatment and does not interfere Se deposition further. Oxygen partial pressure decreases 2.0 – 2.2 times during the reaction, while Se vapors pressure competitively increases. Therefore, inn non-hermetically closed vessel, same effect as for hermetically closed is achieved, while complexity of experimental set-up is significantly decreased.

Considering high ionic contribution to chemical bonding, low thermal conductivity, high non-stoichiometry, polymorphic trend, high melting temperature, halcogenides synthesis by classical methods meets certain experimental difficulties [4].

Therefore, alternative halcogenide synthesis methods, especially with surface modified on nano-level, is of interest both from fundamental, and applied side.

In this paper the method of thermal treatment in selenide vapors is suggested. Applied to thin film coatings, this method allows creating nanometer-size selenide structures integrated to coating and chemically bonded to it.

Thermo-Electric Treatment of Zinc

Zinc films of ~1000 nm thickness were obtained by vacuum thermal evaporation method on glass substrates (Figure 1).

Figure 1. Surface morphology of zinc films obtained by thermal evaporation onto glass substrate.

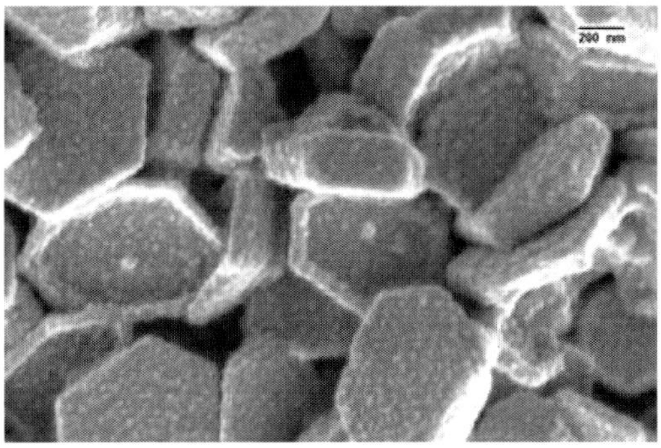

Figure 2. Zinc crystals after treatment with selenide vapors.

For the majority of crystals, hexagonal structure is typical, which is indicative for rather high purity of grown Zn crystals and conditions of either homogeneous, either heterogeneous conditions of crystalline nucleation.

Cross-sectional Zn crystalline dimensions range from 200 to 1000 nm for films of nearly 1000 nm thickness.

Zinc crystals were heated to 250°C for 2 hours in the presence of selenide powder in non-hermetically closed vessel by method analogous to [6].

As a result of treatment in Se vapors, Zn crystalline substrate was coated by nanometer (~20-50 nm) size particles (Figure 2).

Comparison of morphologies of untreated (Figure 1) and treated (Figure 2) in Se vapors Zn films shows that at treated film surface morphological features of a smaller size (20 – 50 nm) are formed, which were not present at initial film.

Also, at the edges of hexahedrons on perimeter protrusions can be noticed. This phenomena can be explained by interacting between Zn and Se vapors.

Another advantageous way of surface modification is developing by us method, where simultaneous temperature and electrical gradient in air atmosphere results in promoting electro-adhesion effects and creating films with significant differences in structure from ones synthesized without electrical field.

Zn film on glass was modified by constant electrical potential U = 300 V, exposition time was 10 min and temperature 250°C. Positive potential was applied to Zn. Precursor sample was placed in a holder that at the same time was upper electrode. Simultaneous effect of thermal treatment and electric field was studied as in [7, 8]. Control sample was placed near treated sample, without applied voltage.

The influence of temperature and electric field on adhesion and phase composition of metallic zinc films was studied by experimental setup shown in Figure 3. Experiments were performed in dry air atmosphere. Films for deposition were deposited on glass substrates.

Treating sample with zinc film was hold in a device serving upper electrode at the same time. Thermal treatment with simultaneous action of electric field was studied. Control sample (outside of the electric field) was placed nearby. Modification of film at glass substrate was handled at $U = 300$ V, treatment time was $t = 10$ min and $T = 250^{\circ}$C.

Obtained results show increased surface resistance ($R_1 \approx 15$ Ohm/sq) comparing to untreated sample ($R_2 \approx 4$ Ohm/sq). This is probably due to Zn oxidation by oxygen ions electro-migrating from glass to Zn. Transition from not-transparency to semi-transparency confirms this assumption. As known, ZnO are optically semi-transparent in visible band.

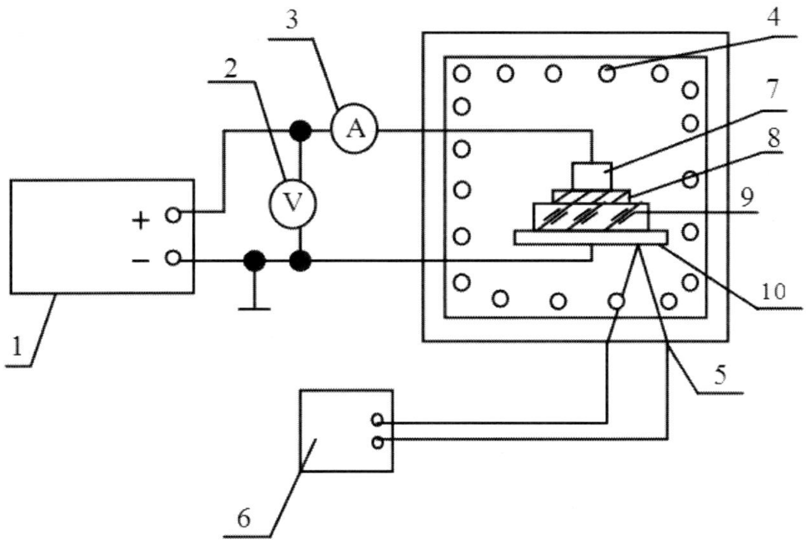

Figure 3. Experimental setup for electro-adhesion experiments. 1 – power supply. 2 – voltmeter, 3 – current recorder. 4. Oven, 5 – thermo-couple, 6 – high sensitive potential meter, 7, 10 electrodes, 8 – substrate for deposition, 9 – dielectric plate.

As a result of complex (temperature-electric) treatment, partial Zn oxidation to ZnO occurs. Large-gap (3.37 eV) semiconductor is a high-ohmic semiconductor with resistivity ~10^{-3} Ohm·cm) at room temperature. Partial Zn oxidation at applied experimental conditions results therefore in a significant decrease in conductivity, which is for pure Zn ~$5{,}9 \cdot 10^{-6}$ Ohm·cm).

Surface morphology was studied by scanning electron microscopy, while surface composition was evaluated by elemental analysis. Those measurements were carried out on Zeiss Merlin) electronic microscope.

Figure 4 shows microphotographs of films surfaces obtained at different stages of treatment.

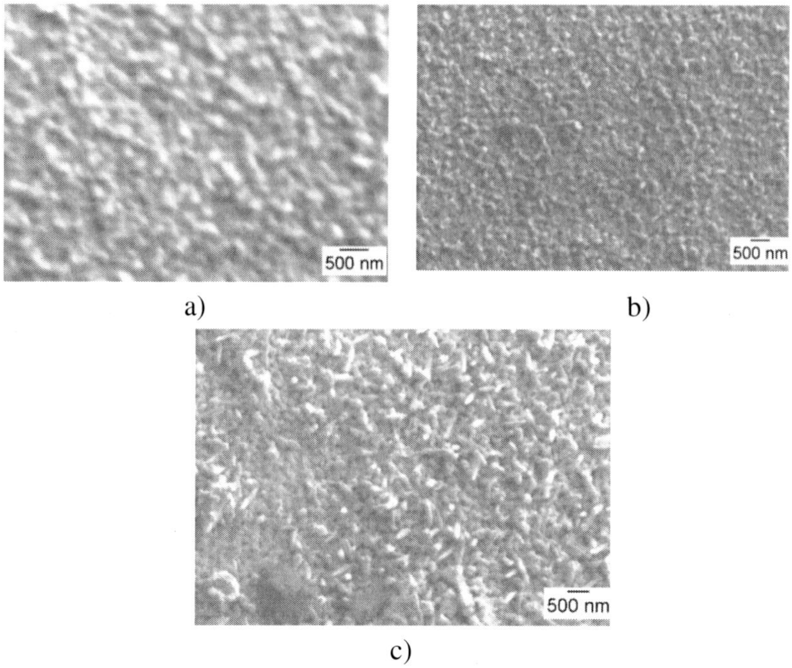

Figure 4. Treated and untreated films SEM images. a) – precursor film, b) – film affected by thermal treatment in air, c) films thermally treated with applied cross-sectional electric field. Exposure time 10 min, temperature 250°C.

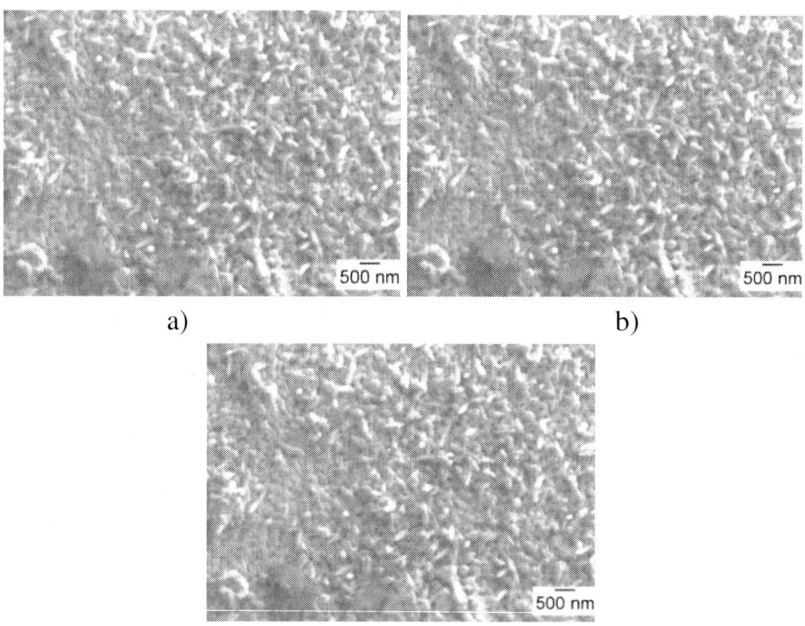

Figure 5. Treated and untreated films SEM images a) precursor film, b) film affected by thermal treatment in air, c) films thermally treated with applied cross-sectional electric field. Exposure time 10 min, temperature 250 C°.

Figure 5a shows that surface consists of separate, adjacent to each other crystals, which size ranges from 50 to 200 nm. From Figure 5b it is clear that temperature does not strongly affect film structure, resulting only with its partial oxidation. Fragment of Zn film surface after 10 min temperature treatment in electric field. Figure 5c is featured by large amount weakly oriented towards surface one-dimensional crystals. While cross-sectional size of those crystals is 50...100 nm, their length is 500... 1000 nm. The majority of filament-shape crystals are oriented towards surface randomly. Therefore, electric field significantly changes film structure: non-uniform electric field favors obtaining ZnO films with developed surface. Those ZnO films are composed from one-dimensional crystals of nanometer size.

CONCLUSION

Preliminary studies have shown that by varying technological regimes of thermal treatment under constant electric voltage, it is possible to alter concentration and linear dimensions of linear zinc oxide crystals.

Metal was converted to the semiconductor by treating in the presence of oxygen and an electric field.

Technological findings obtained in this study are believed to be useful for micro fabrication appliances, where expensive vacuum technologies need to be replaced by more cost-effective surface treatment methods.

REFERENCES

[1] Panchal C. J., Opanasyuk A. S., Kosyak V. V., Desai M. S., Protsenko I. Yu. Structural and sub structural properties of the zinc and cadmium chalcogenides thin films. *J. Nano- Electron. Phys*. 3 (2011) No1, P. 274-301.

[2] Dubey J. P., Tiwari R. K., Upadhyaya K. S., Pandey P. K. Crystal dynamics of zinc chalcogenides I: an application to ZnS. *Turk J Phys* (2015) 39: 242 – 253, http://journals.tubitak.gov.tr/physics/.

[3] Dubey J. P., Tiwari R. K., Upadhyaya K. S., Pandey P. K. Crystal dynamics of zinc chalcogenides II: An application to ZnSe. *IOSR Journal of Applied Physics* (IOSR-JAP) e-ISSN: 2278-4861.Volume 7, Issue 5 Ver. I (Sep. - Oct. 2015), PP 67-75 www.iosrjournals

[4] Triboulet R. *Growth of Zinc Chalcogenides/In book: Crystal Growth Technology*, 2003, pp.497-523.

[5] Mironov M. P., Markov V. F., Mascaevfa L. N., Diakov V. F., Mukhamediarov R. D., Muhamedzianov H. N., Smirnova Z. I., *Method of sensibilization of chemically precipitated lead selenide films for IR radiation*, Patent RU 2357321, (2009).

[6] *JCPDS database No. 03-065-5973.*

[7] Pshchelko N. S., Sevryugina M. P. Modeling of physical and chemical processes of anodic bonding technology. *Advanced Materials Research,* Vol. 1040, 2014, p 513-518

[8] Pshchelko N. S., Tomaev V. V. Polarization of Surface Layers of Ionic Dielectrics at the Interface Between the Electroadhesive Contact and a Dielectric. *Glass Physics and Chemistry*. Vol. 42 No.1. 2016 . P.105 -109.

In: Applied Aspects of Nano-Physics …
Editors: Kirill Levine and Andrey Syrkov

ISBN: 978-1-53614-708-7

Chapter 24

INFLUENCE OF THE SUBSTRATE ON THE FORMATION OF AGI FILMS PREPARED BY LASER ABLATION

***Vladimir V. Tomaev*[1,2,*] *and Tatiana V. Stoyanova*[1]**
[1]Saint Petersburg Mining University, Saint Petersburg, Russia
[2]Institute of Chemistry, Saint-Petersburg State University,
Saint Petersburg, Russia

ABSTRACT

AgI films with 1 μm thickness were synthesized by laser ablation method on substrates of different materials: cover glass, fiberglass and copper foil. The morphology of the surface of all silver iodide films was studied by scanning electron microscopy (measured with the Zeiss Merlin microscope), and their elemental composition was investigated by micro-X-ray spectral analysis. It was shown, that the morphology and composition of the substrate cause a significant effect on the formation of the morphology of AgI films.

Keywords: silver iodide films, laser ablation method, substrates of cover glass, fiberglass and copper foil

---

* Corresponding Author's Email: tvaza@mail.ru.

Introduction

To create a synthesis technology, it is very important to understand the mechanism of the process and the role of factors that determine the kinetics of the reactions. Since most of the processes are structurally sensitive, the most promising substrates for the development of nanotechnologies for the formation of nanoobjects and their use in model processes are monocrystals whose surface has a controlled structure.

Nevertheless, the formation of nanoobjects on substrates that do not have a definite crystal orientation (cover glass, fiberglass, copper foil) can also be very promising because at the early stage of nucleation of nuclei of a growing phase, the action of the substrate on the formation of nanocrystals of a polycrystalline film can be minimal and at this time "quasi-free" growth may occur.

Experimental Section

Sputtering of the films on copper and synthetic-resin bonded [SRB] paper laminate (getinax) substrates was carried out [1] by laser ablation [2] using an Xe-Cl excimer ultraviolet laser with a radiation wavelength of 30 nm, a pulses energy of 10 mJ, a duration of 30 ns, and repetition rate of 14 Hz. In previous work [3, 4] was shown that this method allows to obtain AgI films with a composition that coincides with the composition of the target. In the process of growth, the shape of the surface (degree of roughness) can play an increasing role and at the final stage of growth the shape of the surface of the film can "copy" it.

It was observed a correlation between the morphology of the surface of the getinax and copper foil substrates with the corresponding morphologies on the surface of the AgI films. The surface of the silver iodide film deposited on the getinax repeats the morphology of the getinax substrate and has an inhomogeneous character. The pore size on the surface in some cases reaches several microns. However, the average size of individual grains is 100-200 nm.

To explain the results obtained, the table gives the coefficients of thermal expansion of all interacting surfaces: AgI; copper foil; Getinax. As can be seen from the table, the coefficient of thermal expansion of hexagonal β-AgI has a negative value for all temperatures at which this phase exists (from 0 to 423 K) [3]. At the same time, the coefficients of thermal expansion of copper and getinax have a positive coefficient of thermal expansion [4, 5].

In this respect, the interaction of a relatively thin hexagonal β-AgI film with a negative coefficient of thermal expansion with thick substrate layers (copper and getinax), each of which has a positive coefficient of thermal expansion, can be very effective, which leads to a change in the surface of the film due to large difference coefficients of thermal expansion. A similar effect was previously observed by the authors on the example of other materials in interaction with β-AgI [6].The analysis of the substrate effect on the surface morphology of the AgI film grown by laser ablation is based on the classical theory of nucleation [7] and the Gibbs-Volmer equation.

Table. Coefficients of thermal expansion of the film material (AgI - silver iodide) and substrates (copper foil and getinax)

Material	Coefficient of thermal expansion (10^{-6} °C^{-1})	Reference
AgI	-8±4	[3]
Cu	16,5	[4]
Getinax	7,9-8,7	[5]

Explanation of the obtained results depending on the properties of the film, substrate and technological conditions is possible with the help of the mechanisms of Wolmer-Weber, Frank van der Merve and Stransky-Krastanov.

CONCLUSION

AgI films were synthesized by laser ablation method using an Xe-Cl excimer ultraviolet laser on substrates of different materials: cover glass, fiberglass and copper foil. The film surfaces were studied with scanning electron microscopy. It was observed a correlation between the morphology of the surface of substrates with the corresponding morphologies of the surface of the AgI films. The analysis of the substrate effect on the surface morphology of the grown films is based on the classical theory of nucleation [7] and the Gibbs-Volmer equation. An explanation of the results obtained, depending on the properties of the film, substrate and technological conditions, is possible with the help of the Wolmer-Weber, Frank- van der Merve and Stransky-Krastanov.

The results of the analysis make it possible to conclude that it is possible to control the degree of nanodispersity of the silver iodide film surface by changing the substrate material and the technological application parameters. Such a modification, as was shown in [7], can lead to structural phase transitions of silver iodide, as well as a change in the grain size in the film, which as a result may lead to an increase in the conductivity with respect to Ag+ ions.

Acknowledgments

This work was partly supported by Russian Science Foundation 17-03-00121.

References

[1] Hull S. Superionics: crystal structures and conduction processes. *Rep. Prog. Phys.* 2004, 67, pp 1233–1314.

[2] Bäuerle D. *Laser Processing and Chemistry*. Berlin, Heidelberg: Springer-Verlag, 2000. p. 788.

[3] Volmer M. and Weber A. Keimbildung in Übersättigten Gebilden [Nucleation in Supersaturated Forms], *Z. Phys. Chem.* Leipzig 1926, pp 119 - 277.

[4] Harvey G. and Fletcher N.H. Thermal expansion of β-silver iodide at low temperatures 1980, *J. Phys. C: Solid St. Phys.*, 13 pp. 2969-2973.

[5] Koshkin N.I., Shirkevich M.G. *Handbook of Elementary Physics*. 1972, M. Science. p. 256.

[6] *Getinaks electrotechnical sheet. Technical conditions*. GOST 2718-74. 1997, M. Publishing House of Standards. p. 24.

[7] Tomaev V.V., Tver'yanovich Y.S., Bal'makov M.D. Methods of control over the phase composition of nanostructured silver iodide *Russian Journal of Electochemistry*, 2017, Volume 53, Issue 7, pp 777–781.

In: Applied Aspects of Nano-Physics …
Editors: Kirill Levine and Andrey Syrkov
ISBN: 978-1-53614-708-7

Chapter 25

CAPABILITIES OF MARNETIC ABRASIVE FINISHING IN THE FIELD OF NANOTECHNOLOGY

E. G. Zlotnikov,* A. I. Keksin and L. E. Koroboyankwe
Saint Petersburg Mining University, Saint Petersburg, Russia

ABSTRACT

This chapter explains the advantages of magnetic abrasive treatment over the traditional methods used in the mechanical process of finishing. The possibility of forming nanorelief surfaces of various components and products is shown. The results of experimental studies on magnetic abrasive finishing of complex profile surfaces for cutting tools are also presented.

Keywords: magnetic abrasive finishing, nanorelief, nanotechnology, finishing methods

GOALS

The aim of this chapter is to show capabilities of magnetic abrasive finishing in the field of nanotechnology.

---

* Corresponding Author's Email: zlotnik_evg@mail.ru.

Introduction

Since the last quarter of the twentieth century, research in the field of nanotechnology [1] has become a priority in scientific research. Nanotechnology refers to technological processes and systems of machines and mechanisms, capable of performing ultra-precise operations on a scale of several nanometers [2]. It is known that micro- and nanorelief of machine components' surface layers are formed at the final stages of the technological process of manufacturing parts, i.e., during the finishing operations.

Currently, traditional methods of finishing (grinding, honing, lapping, etc.) do not allow for the provision of a high-quality level of the surface layer [3], which must be formed on many components in various spheres of human activity: in engineering (for example, in electronics, optics, lasers); in ecology, medicine, etc. So, for example, the roughness quantity provided by honing and grinding on the surface of a component is Ra = 25 nm (0.025 μm), and Ra = 13 nm (0.013 μm) respectively [3], while one of the primary tasks of the electronics industry in the manufacturing of silicon monocrystal plates is to decrease the height of microroughness from 10 nm (0.01 μm) to 1 nm (0.001 μm) [4].

Materials and Methods

Providing such a high-quality level of the components surface layer is possible by creating progressive technologies based on the use of physical and chemical sources of energy. One such source is the magnetic field energy, which is increasingly used in many fields of technology. Based on the use of magnetic field energy, a method of magnetic abrasive finishing has been developed, which makes it possible to form a nanorelief with unique properties [4].

Processing by this method consists of oriented as well as selective abrasive micro-cutting and micro-smoothing. The essence of this process is that, at relatively high values of micro-protrusions, the powder grains are contacted mainly with the shell peaks, which are concentrators of the magnetic field lines. Each working element (grain) in the magnetic field is set by the largest axis towards the surface to be treated. During wear and dulling of vertices, the element is reoriented in such a way that the newly formed largest axis is guided along the magnetic field lines. As a result, the surface of the workpiece is machined with sharp edges, i.e., the process of oriented abrasive cutting [6] takes place. This feature of magnetic abrasive finishing gives huge advantages over traditional methods of finishing operations. For example, magnetic abrasive finishing of silicon monocrystals plates, details of optics and laser technology allows for the formation of nanorelief surfaces with an unevenness height of less than 2 nm [4].

In the Mechanical department laboratory of machines with CNC, in St. Petersburg Mining University, experimental studies were carried out on magnetic abrasive finishing of complex profile surfaces of cutting tools. Machine Taps M16 × 2 made of high-speed steel P6M5 were taken as samples. Used to form a magnetic abrasive powder based on titanium carbide and iron (TiC + Fe) was a magnetic abrasive brush in the workspace of the electromagnetic installation system based on a CNC milling machine.

The best result was obtained during magnetic abrasive finishing with powder granularity $\Delta = 160$ μm, magnetic induction $B = 0.6$ T, polishing time $t = 60$ s. The parameter of the roughness of the rear surfaces of the cutting edges of the taps was achieved within Ra = 0.061 ... 0.09 μm (61 ... 90 nm) with the initial value Ra = 0.16 ... 0.22 μm [7, 8].

Results and Discussion

With the use of magnetically abrasive powders of less granularity, the selection of the regime parameters and the increase in the finishing time of the surfaces of the cutting tools, it is possible to obtain a nanorelief with a roughening parameter of 40 ... 20 nm. Advantages of the method of magnetic abrasive polishing include the formation of a surface layer with favorable compressive stresses and increased microhardness, a significant reduction in the number of surface defects, high productivity and the possibility of process automation.

Conclusion

In conclusion, it can be said that magnetic abrasive finishing has a tremendous technological potential. The unique combination of the combined action of magnetic field pulses on the material structure and the "soft" modes of removing the defective surface layer makes it possible to form a surface with a minimum of defects and high-performance properties, such as resistance to corrosion, wear and mechanical stress, by a magnetically abrasive method.

References

[1] Shevchenko L.B. *Information resources nanotechnology.* URL: http://cyberleninka.ru/article/n/informatsionnye-resursy-po-nanotehnologiyam (accessed date: 31.10.2016). (in Russian).

[2] Taniguchi N. Current status in, and future trends of ultraprecision machining and ultrafine material processing. *Annals of CIRP*, Vol. 32/2, 1983, P. 573-582.

[3] Sunil Jha V.K. Jain. *Nano-finishing techniques*. URL: http://web.iitd.ac.in/~suniljha/nanofinishing.pdf (accessed date: 31.10.2016).

[4] Khomich N.S. *Magnetic-abrasive machining of the manufactured articles*. Minsk: BNTU, 2006. – 218 p. (in Russian).

[5] Kremen G.Z. Machining time estimation for magnetic abrasive processes. *International Journal of Production Research*, Vol. 32, 12, 1994, 2817-2825.

[6] Akulovich L.M. *Fundamentals of magnetic abrasive finishing of metal surfaces*. Minsk: BGATU, 2012. – 316 p. (in Russian).

[7] Keksin A.I. Methods of increasing the quality of the thread pitches / A.I. Keksin, V.V. Maksarov. *Agronomy Research*. 2013. – Vol. 11, 1. – P.139-146.

[8] Keksin A.I. Technological improvement of the quality of complex surfaces by the method of magnetic abrasive polishing. A.I. Keksin, V.V. Maksarov. *Metalloobrabotka*. 2017. - 1 (97). - P. 47-57. (in Russian).

In: Applied Aspects of Nano-Physics …
Editors: Kirill Levine and Andrey Syrkov
ISBN: 978-1-53614-708-7

Chapter 26

REACTIVITY OF SOLID SURFACES IN THE CHEMICAL NANOTECHNOLOGY

Yu. K. Ezhovskii
Saint Petersburg State Technological Institute (Technical University),
Saint Petersburg, Russia

ABSTRACT

The concept of the macromolecular structure of solids and quantitative estimation of surface reactivity using inductive constants is applied to the estimate of surface nanostructures on dispersed and single crystal matrices. The possibilities of the proposed approach are demonstrated by the example of the synthesis of oxide nanostructures on a silicon surface.

Keywords: reactivity of solid surfaces, dispersed and single crystal matrices, oxide nanostructures

INTRODUCTION

The development of the chemical nanotechnology of low dimensional systems on a solid surface are mainly associated with the possibility of predicting the conditions of reactions using quantitative estimation of their reactivity. One such approach is a correlation analysis system based on the principle of the linearity of free energies and is successfully used in the chemistry of high molecular weight compounds [1]. According to the core models of the structure of solid materials and the chemistry of supramolecular compounds developed on its basis [2, 3], any chemical transformations on a solid surface can be considered as similar to those of a polymer. In this case, an approach based on application of the principle of the linearity of free energies and a correlation-analysis system in the chemistry of a solid surface can become the basis for the creation of a quantitative theory of surface reactions.

EXPERIMENTAL SECTION

In accordance with the core concept of the structure of solids [2], any solid compound is a combination of a chemically inert core (*X*–) and reactive functional groups of the surface (*Y*) (e.g., –OH). All surface reactions with the participation of these groups not affecting the structure of the core can be considered as similar to polymer transformations [3]. The ability of the core as a substituent to any type of interaction is quantitatively characterized by the corresponding constant (σ_X), the numerical value of which has a relative character, i.e., makes it possible to estimate the effect of the given substituent with respect to the reference one. This makes it possible to apply the principle of the linearity of free energies for different solid compounds having similar-type functional groups within the formal approach and use linear correlations of the type:

$$A = A_{XY} + \alpha_Y \sigma_X, \tag{1}$$

where A is any correlated quantity (e.g., the velocity constant (K) or log$K$); α - is the proportionality coefficient reflecting the sensitivity of the reaction center to the effect of the substituent in the given reaction. Indices Y and X refer, respectively, to the reaction center (the functional group) and the solid core.

The influence of the core of the solid, and, more exactly, the surface group of its atoms can be characterized by the value of the inductive constant (σ_S), which can be determined, e.g., from shifts of the frequencies of valence vibrations of the element–hydride bond, in particular, the Si–H bond in the functional group of the surface compound. This bond is very sensitive to the effect of substituents and the shift of its frequencies almost excludes the effect of steric factors. In this case Eq. (1) takes the form:

$$\nu_i = \nu_0 + \alpha\,(\Sigma\,\sigma_i + \sigma_\Sigma), \tag{2}$$

where ν_i, ν_0 are the frequencies of the valence vibrations in the given compound and in the reference one, e.g., methyl-substituted silane or siloxane.

Results and Discussion

The coefficients of Eq. (2) for a reaction series of solid oxides were determined using samples of silicon dioxide (aerosil A-175) with different chemosorbed chlorinated silanes. As a result of mathematical processing of the experimental data (ν_i) obtained for a series of surface compounds and values of Taft constants ( ) for such substituents as Cl_2MeSiH ($\Sigma\sigma_i^* = 2.88$), Cl_3SiH ($\Sigma\sigma_i^* = 5.76$), Cl_2PhSiH ($\Sigma\sigma_i^* = 3.48$) and $ClPh_2SiH$ (($\Sigma\sigma_i^* = 1.10$), we determined the coefficients of the relation (2) and obtained the expression:

$$\nu\iota = 2064 + 19.21\ (\Sigma\ \sigma\iota^* + \sigma^*\Sigma), \tag{3}$$

Analysis of experimental results on determining the reactivity of hydroxyl groups of silica, silicon and arsenide–gallium matrices made it possible to assume that the hydroxyl groups on the real surface of a semiconductor matrix will possess an activity determined by the prehistory of the sample, namely, the thickness of the oxide layer formed after preliminary treatment. It is possible to determine this influence from the inductive effect and reveal the quantitative regularity of the reactivity of OH-groups of the surface of the crystal on the thickness of the oxide layer. The obtained data made it possible to establish the quantitative relationship between the thickness of the oxide layer on the silicon matrix and its inductive effect, and, consequently, the activity of hydroxyl groups. The values of the inductive constants make it possible to quantitatively estimate the effect of the solid core and the state of its surface at different temperatures, predict the direction of surface reactions with different reagents and establish their stoichiometry, when the use of direct chemical-analytical methods is difficult.

References

[1] Pal'm, V. A. *Bases of Quantitative Theory of Organic Reactions* (Khimiya, Leningrad, 1977) [in Russian].

[2] Aleskovskii, V. B. *Zh. Prikl. Khim.* 55, (1982). 725-730.

[3] Aleskovskii, V. B. *Course of Chemistry of Supramolecular Compounds* (SPb. Gos. Univ., St. Petersburg, 1996) [in Russian].

[4] Ezhovskii, Yu. K. *Russ. Chem. Rev.* 73, (2004). 195- 204.

In: Applied Aspects of Nano-Physics …
Editors: Kirill Levine and Andrey Syrkov

ISBN: 978-1-53614-708-7

Chapter 27

THE INFLUENCE OF THE MATERIAL STRUCTURE OF AN INSTRUMENT ON THE QUALITY OF PROCESSING WHILE MILLING WITH CUTTING CERAMICS

A. D. Khalimonenko*[*], *R. R. Rakhmankulov and I. V. Gorshkov
Saint Petersburg Mining University, Saint Petersburg, Russia

ABSTRACT

The paper deals with the problems that arise when milling the mounting surfaces with a tool equipped with cutting ceramics, based on an analysis of the requirements applicable to the quality of the surface layer and modern technologies for its providing. The results of the research showed that the quality improvement in the case of finishing milling depends on the operating capacity of the ceramic tool, determined by the parameters of its microstructure.

Keywords: cutting ceramics, operating capacity of tool, milling, processing quality, mounting surface

[*] Corresponding Author's Email: khalim76@rambler.ru.

GOALS

The purpose of this work is to determine the operating capacity of a ceramic tool that affects the quality of machining while milling the precise elements of machine parts. The operating capacity of the ceramic material is determined on the basis of its microstructural parameters.

INTRODUCTION

A feature of modern production is the use of new technologies and tool materials, which, having high cutting properties, ensure, at the same time, the necessary quality of processing of the precise elements of machine parts. Such precise machine parts include elements of the mounting surfaces, which are subject to relatively high requirements to ensure the necessary dimensional accuracy, surface roughness, the accuracy requirements for the geometrical elements and their mutual disposition.

The most optimal is the technology: roughing, semi-finished and fine (fine) milling, the latter being recommended for use with cutters equipped with replaceable plates of cutting ceramics, which provide the necessary quality of processing at the highest productivity. This technology has a disadvantage - low stability of cutting ceramics with intermittent cutting, leading to early failure and loss of efficiency of not only a separate cutting insert of the mill, but also the entire tool.

If one of the ceramic plates fails to work, the load on the remaining plates increases, the tool quickly fails, which immediately leads to loss of processing quality. Hence it follows that the task of this study is to determine the operability of each ceramic plate in order to install in the milling cutters of such ceramic plates whose performance is at the same level.

MATERIALS AND METHODS

Cutting ceramics differs from other materials with high strength, reduced brittleness and tendency to cracking, while possessing high cutting properties [1]. The operating capacity of the cutting ceramics is influenced by such microstructure parameters as the amount of carbide grains and their average diameter. Ceramic plates having a finer grain structure (average grain size of 1.2 ... 1.5 μm) and a smaller percentage of porosity (about 8%) have greater operating capacity compared to ceramic plates in which the structure consists of large-diameter grains (medium grain size 1.9 ... 2.5 μm) and there is a relatively large percentage of porosity (about 12%) [1, 2].

The selection of ceramic plates for milling, according to their performance, was carried out by the correlation dependence of the microstructural parameters and the specific electrical resistance of the tool material. Ceramic plates with a larger microstructure have values of resistivity close to R = 10 Ω, and ceramic plates with a smaller microstructure have electrical resistivity parameters close to R = 100 Ω [1]. A similar range of parameters of the electrical resistivity of ceramic plates does allow them to be in the same group in terms of their microstructural parameters and in terms of operating capacity [1, 3].

For the experiments, the end mill was used and the recommended finishing modes adopted for the standard technology for the preparation of mounting surfaces were selected.

RESULTS AND DISCUSSION

A series of experiments was conducted in which a milling cutter equipped with ceramic plates with similar electrical resistivity parameters (a group of plates with R = 75 ... 100 Ω) showed the following results of the milling process stability: the actual tool path without loss of processing quality was l = 32 m, after which there was a sharp deterioration in the roughness of the surface being treated and going beyond the tolerance range of the machined element, the actual period of tool life, when it could work without loss of quality was T = 35.6 min.

The examination of the treated surface showed that before the instrument reached the maximum value of operability, the quality of the treatment was within a range sufficient for details of this type - the surface roughness was not lower than Ra = 1.6 μm, the flatness deviations did not exceed 0.03 mm for a length of 100 mm.

In the second series of experiments, one of the milling cutter plates was chosen in such a way that its efficiency and specific electrical resistance differed sharply among the representatives of the group (R = 10 Ω). The tests carried out showed that the path traveled by the tool without loss of processing quality and the actual period of durability decreased by 1.5 times.

CONCLUSION

Thus, it can be concluded that equipping the cutter with replaceable ceramic plates with values of specific electrical resistance belonging to one group allows not only stabilize the milling process, but also significantly improve the tool's efficiency by increasing the period of its durability, which will ensure the required quality of the treated surfaces blanks of machine parts. On this basis, it is recommended to equip the multiblade cutting tool with such ceramic plates, performance criteria which are in the same group.

In order to increase the implementation life of the tool equipped with cutting ceramics, to increase the processing capacity and to ensure the required quality indicators, it is desirable to mount such ceramic plates in the tool, the values of the resistivity of which approach R = 100 Ω.

REFERENCES

[1] Maksarov V., Khalimonenko A., Timofeev D. Machining quality when lathing blanks with ceramic cutting tools. *Agronomy Research*, Tartu: Estonian University of Life Sciences, 2014, Vol. 12, P. 269-278.

[2] Maksarov V., Khalimonenko A., Olt J. Effect of porosity on the performance of cutting ceramics. *Agronomy Research*, Tartu: Estonian University of Life Sciences, 2016, Vol. 14, SI 1, P. 1043-1052.

[3] Maksarov V., Khalimonenko A. *Forecasting performance of ceramic cutting tool. Key Engineering Materials,* Trans Tech Publications, 2017, Vol. 736, P. 86-90.

In: Applied Aspects of Nano-Physics …
Editors: Kirill Levine and Andrey Syrkov

ISBN: 978-1-53614-708-7

Chapter 28

INVESTIGATION OF WEAR-RESISTANCE OF THIN-LAYER CARBIDE-TITANIUM COATINGS OBTAINED BY VACUUM-ARC (PVD) DEPOSIT

Viktor A. Krasnyy [*] ***and Vyacheslav V. Maksarov***
Saint Petersburg Mining University, Saint Petersburg, Russia

ABSTRACT

A review of studies on the use of coatings from titanium carbide has been carried out. Examples are given of the use of TiC coatings to increase the wear resistance of a number of critical parts, and also as an anticorrosion coating in various fields: aerospace, nuclear power engineering and automotive industry. The use of titanium carbide for solving a number of tribotechnical problems is considered. The peculiarities of the process of vacuum-arc deposition of TiC coatings are studied. The influence of a change in the technological regimes of vacuum arc deposition on changes in the density of coatings, as well as in the ratio between titanium carbide and individual elements is shown. The problems of optimization of TiC coatings obtained by the PVD method for wear resistance with regard to friction at high pressures with lubrication are considered. Studies have been carried out to study the effect of technological regimes of pressure variation in the working chamber and the potential applied to the substrate with a specific wear rate during friction of two rotating cylinders with lubrication.The technological modes of PVD for the formation of optimal for wear resistance of TiC coatings.

Keywords: wear resistance, carbide-titanium coatings, vacuum arc spraying, specific wear rate

[*] Corresponding Author's Email: vikras1955@yandex.ru.

Goals

The aim of the work was to study the influence of vacuum-arc spraying technological modes on the wear resistance of carbide-titanium coatings by changing the pressure in the working chamber and the potential applied to the substrate, with respect to friction at high pressures with lubrication.

Introduction

Titanium compounds have found wide application in various industries. This is due to the fact that it has a number of unique properties, such as super hardness, high heat resistance, high modulus of elasticity, resistance to acids and alkalis, good electrical conductivity.

Coatings of titanium carbide are widely used and applied in various fields: aerospace, nuclear power, automotive. TiC is one of the most effective materials used as a wear-resistant coating. New prospects for the use of titanium carbide are discovered when using it in a nanostate (modification of alloys of different composition and purpose).

For the deposition of high-quality titanium-based coatings, the vacuum-arc deposition method (cathode-arc deposition) is a widely used method of coating thin films in a vacuum by condensing material from plasma streams generated at the cathode - a target in the cathode spot of the vacuum arc of a high-current low-voltage discharge developing exclusively in pairs of electrode material. The method is also known under the names: cathodic arc deposition (Arc-PVD), CIB-cathodic-ion bombardment method. The method of vacuum-arc spraying allows a wide variation in the structure of the emerging coatings due to a change in the pressure in the working chamber and the potential applied to the substrate. The change of technological regimes leads to changes in the density of coatings, as well as in the ratio between titanium carbide and individual elements. Obviously, such structural transformations should influence the wear resistance of the coatings obtained.

Materials and Methods

Experimental studies were carried out on a standard friction machine. The investigated stationary sample coated with TiC-roll is made of steel 20 and had a diameter of 46 mm. The rotating roller with a diameter of 46 mm was uncoated and was made of steel 45 followed by quenching (42 ... 45 HRC). The coating was formed at two pressure levels C_6H_6 - 0.17 Pa and 0.27 Pa and four potential levels on the substrate - 50,

100, 150 and 200 V. The coating thickness was 6-10 μm. To lubricate the friction unit, an oil bath with a volume of 200 cm^3 was used, into which a rotating roller was partially immersed (2 mm). As the lubricating medium used industrial oil. Samples of 40Cr steel without quenching (28 ... 32 HRC) and quenching (42 ... 45 HRC) were also tested to assess the comparative wear resistance of coatings at high pressures.

RESULTS AND DISCUSSION

The results of comparative tests of the wear resistance of chromium steel and TiC coatings applied at pressures of C_6H_6 -0.17 Pa and 0.27 Pa at stresses in the 200 V substrate showed that the specific wear rate for coatings in 6 ... 10 times lower than for hardened steel. In addition, it can be concluded that the wear resistance of TiC coatings depends strongly on the pressure in the chamber when coating is applied. The coating applied at a higher pressure has a significantly higher wear resistance. In this connection, for subsequent tests on the influence of the potential of the electrode, the samples were made at a chamber pressure of 0.27 Pa. At a load of 500 N compared with a load of 50 N, an intensive coverage of the coating is observed, but a significant difference in the wear resistance of coatings applied at stresses on a 50-100 and 150-200 V substrate is observed. The minimum value of the specific wear rate was attained at a voltage of 150 V, and the critical pressure was noticeably higher for TiC coatings formed at substrate potentials of 150 and 200 V and reached a value of the order of 110 MPa.

CONCLUSION

As a result of the performed studies, technological regimes of VDN were chosen to form optimal TiC coatings for wear resistance, some features of wear of such coatings at large and small loads under lubrication conditions were considered. The data obtained make it possible to expand the fields of application of TiC coatings to solve complex tribological problems.

REFERENCES

[1] Shiryaeva L.S., Garbuzova A.K., Galevsky G.V. Production and application of titanium carbide (assessment, trends, forecasts). *Scientific and technical lists of the St. Petersburg State Polytechnic University*. 2014, № 2 (195), pp. 100-108.

[2] Burov I.V, Lisenkov A.A. Current state and prospects for the development of vacuum arc welding technology for wear-resistant coatings. *Vacuum technical and technology*. 2002. Vol. 12, No. 1. pp. 55-60.

[3] Krasnyy V., Maksarov V., Olt J.Improving fretting resistance of heavily loaded friction machine parts using a modified polymer composition. *AgronomyResearch.* 2016, 14(S1), pp. 1023-1033.

[4] Maksarov V.V., Krasnyy V.A. The mechanisms of friction of thin-layer nano-coatingsunder conditions of fretting. *Scientific and technical lists of the St. Petersburg State Polytechnic University*. 226, (2015),№3, pp. 111-120.

In: Applied Aspects of Nano-Physics ...
Editors: Kirill Levine and Andrey Syrkov

ISBN: 978-1-53614-708-7

Chapter 29

FORMATION OF HYDROPHOBIC COATING ON THE MODIFIED POLYMERS SURFACE

Valeria Semenova* and Anna Kochetkova
Saint Petersburg State Institute of Technology,
Saint Petersburg, Russia

ABSTRACT

Chemical surface modification of polyethylene (PE) by phosphorus oxide (-P-O-) and phosphorus-silicon-organic (-P-O-Si-R) structures was carried out by the method of molecular layering (ML). The degree of hydration of the modified samples surface in conditions of storage in air atmosphere with a constant relative humidity was controlled by using the method of atomic force microscopy (AFM). The effect of imparted functional groups in the degree of hydrophobicity of the PE surface was assessed by using the method of measuring the contact angle. The study on the presence of physicaly sorbed water on the surface of the modified PE samples, which were storing in a desiccator with a relative humidity of 85%, was carried out by the method of IR-spectroscopy.

Keywords: polyethylene, molecular layering, atomic force microscopy.

INTRODUCTION

Low cost and availability of PE make it a perspective material for the creation of film electrets. Nowadays polymeric electrets as sources of a constant electric field are widely

* Corresponding Author's Email: semenovaleria@mail.ru.

used in various fields of science and technology. Due to the fact that the electrets are based on pure polymeric materials, have not the best properties, it's necessary to modify the film surface by structures that are able to amplify polarization effects. It is known that the formation of nanoscale -P-O-structures on the material surface leads to appearance of long-stored electret properties in polymer [1]. However, in the process of store -P-O-groups, having a high hydrophilicity, form on the film surface the layer of sorbed water from the air. It leads to rapid degradation of electret properties and makes searching the ways of hydrophobic surface modification of polymers the actual task.

MATERIALS AND METHODS

The objects of study were samples of PE film with modified surface by -P-O- and -P-O-Si-R structures, which were storing after the synthesis in desiccators with different relative humidity (50% and 85%), to determine the influence of relative humidity on the rate of hydration of the surface of the studied samples.

Synthesis of -P-O structures on the surface of PE was carried out according to the method of molecular layering [2] at temperature of 60°C by an alternate treatment of PE by pairs of $POCl_3$ and by followed vapor-phase hydrolysis. To make the surface of the modified polymer hydrophobic, after synthesis -P-O-structures the material was treated in pairs of $Si(CH_3)_2Cl_2$ under the same conditions with followed vapor-phase hydrolysis of unreacted chlorine atoms.

The investigation of samples surface morphology was carried out using scanning probe microscope Solver P47 Pro (NT-MDT, Russia) by the method of AFM in tapping mode.

RESULTS AND DISCUSSION

According to AFM research the total overlap of the sample surface with -P-O-groups by layer of sorbed water when stored in a desiccator with a relative humidity of 50% happens after 35 days. For the sample with a double system, the process under the same storage conditions takes about 2 months. According to the data of AFM research whole flooding of sample surface with -P-O-structures in conditions of 85% relative humidity occurs after ~ 21 days of storage, while in the case of the sample with a double system, the process of full hydration of the surface takes almost 3 times longer.

Thus, it was shown that Si-R-groups block -P-O-structures and greatly slow down the sorption of water from the environment on the surface of the polymeric composite. It should help to extend the operation of the electret material.

Acknowledgments

Research was partly financed from Ministry of Education and Science of Russian Federation (project No 16.1798.2017/4.6)

References

[1] Rychkov A.A., Rychkov D.A., Trifonov S.A. Stability of electret charge in the films and fiber composite polyethylene modified by pairs of phosphorus trichloride. *Scientific Journal of Herzen University*. V.8, No. 38. 2007. P. 89 - 97.

[2] Malygin A.A. From chemical reactions at the solid surface to molecular layering nanotechnology. *Bulletin of Saint Petersburg State Institute of Technology* (Technical University). No.1 (27). 2007. P. 14 - 24.

In: Applied Aspects of Nano-Physics …
Editors: Kirill Levine and Andrey Syrkov
ISBN: 978-1-53614-708-7

Chapter 30

ANTIFRICTION PROPERTIES OF NANOCOMPOSITE MATERIALS BASED ON ALUMINUM REINFORCED WITH CARBON NANOFIBERS FOR APPLICATION IN FRICTION COUPLES OPERATED UNDER EXTREME CONDITIONS

Aleksandra N. Skvortsova*[*], *Aleksander D. Breki*, *Oleg V. Tolochko*, *Tatiana S. Koltsova* and *Sergey E. Aleksandrov

Peter the Great Saint Petersburg Polytechnic University,
Saint Petersburg, Russia

ABSTRACT

Research in the field of creation of materials from metals reinforced by carbon nanostructures in the last decade is growing rapidly. As a rule, for the introduction of carbon nanotubes and nanofibers (lip) in a metal matrix (in particular, aluminum) is its hardening, improvement of wear resistance and improved anti-friction properties. The main problem associated with the manufacture of composite materials is the agglomeration of nanotubes due to their high surface energy, which can degrade the expected properties as antifriction, and antiwear. The author of this work has been considered a method of synthesis of carbon nanofibers directly on the aluminum microparticles, allowing to obtain a uniform distribution of lip in the aluminum matrix.

[*] Corresponding Author's Email: annelet@yandex.ru.

This chapter presents the results of a comparative laboratory study of antifriction properties of composite materials based on aluminum reinforced with carbon nanofibers and the friction for 12H steel. The dependences of the frictional force on time for the developed composite materials and matrices implemented a comparison of the results obtained for composite materials and pure aluminum samples.

Keywords: sliding friction, friction properties, carbon nanofibers, nanocomposite material, aluminum

Description of the Method

It is widely known that the task of creation of lightweight materials with improved antifriction properties is an urgent challenge in innovative economy. In this regard research in the field of creating metal materials reinforced with carbon nanostructures has been actively developing over the last decade [1].

The goals of incorporation of carbon nanotubes and nanofibers (CNF) into a metal matrix are its hardening, increasing wear resistance and improving antifriction properties. The main problem associated with the manufacturing of composite materials is the agglomeration of nanotubes due to their high surface energy. This phenomenon often leads to deterioration of expected antifriction and anti-wear properties.

The authors in [2] propose novel method of synthesizing carbon nanofibers directly on aluminum microparticles. It provides uniform distribution of CNF in the volume of aluminum matrix.

Results and Discussion

The results showed that the aluminum samples obtained at different technological parameters show different friction properties in the mode of a translational sliding friction on a plane of 12H steel. Antifriction action of carbon nanofibers is shown in the moment of breakaway and during the movement of the friction surfaces relative to each other, indicating the influence of the lip on the properties of oxide films of specific composite materials. The most stable friction coefficients were detected in the composite material ALM18+1.5% of the lip. This material was observed, with the minimum value of the steady-state friction force being (≈ 6H) 25% less than ALM16, and 62.5% (2.7 times) less than the ALM102 material with greater hardness.

REFERENCES

[1] Skvorcova, A. N., K. A. Lycheva, A. A. Voznyakovskij, T. S. Koltcova. Kompozicionnye materialy na osnove alyuminiya, uprochnyonnye uglerodnymi nanovoloknami [Aluminum-based composite materials hardened with carbon nanofibers]. *Nauchnotekhnicheskie vedomosti SPbGPU* [*Scientific and technical statements SPbGPU*]. 2015. № 3(226). S. 78-82.

[2] Rudskoj, A. I., O. V. Tolochko, T. S. Koltcova, A. G. Nasibulin. Sintez uglerodnyh nanovolokon na poverhnosti chastic poroshka alyuminiya [Synthesis of carbon nanofibers on the surface of plastic powders of aluminum]. *Metallovedenie i termicheskaya obrabotka*. 2013. №10. S. 53-57.

In: Applied Aspects of Nano-Physics …
Editors: Kirill Levine and Andrey Syrkov
ISBN: 978-1-53614-708-7

Chapter 31

FEATURES OF FORMATION OF ZINC NANOPARTICLES ON SUBSTRATES OF GLASS AND QUARTZ

***Vladimir V. Tomaev*[1,*] *and Vladimir A. Polishchuk*[2,3]**

[1] Saint Petersburg Mining University, Saint Petersburg, Russia
[2]University of Information Technologies,
Mechanics and Optics, Saint Petersburg, Russia
[3]Admiral Makarov State University of Martitime and Inland Shipping,
Saint Petersburg, Russia

ABSTRACT

Zinc films obtained by magnetron sputtering on glass, and quartz substrates are investigated. The morphology of the surface, the thickness of the films, and the size of the zinc nanocrystals were studied by scanning electron microscopy. It is established that the morphology of the surface of zinc films essentially depends both on the technological application conditions and on the substrate material. Zn film forms very small islands on initial stages of their growth on quartz substrate, but forms more significant islands on glass substrates. The geometric and thermodynamic parameters of the investigated films were estimated. Resistance of zinc films was investigated by impedance spectroscopy. The conclusion is drawn that the conductivity of the films is of a percolation nature. As film thickness increase, so does the twinning of Zn crystals.

Keywords: zinc films, magnetron sputtering, glass and quartz substrates

*Corresponding Author's Email: tvaza@mail.ru.

Introduction

This chapter reports some results relating to the thin Zn films formed by magnetron sputtering on glass and quartz substrates. The problems of obtaining thin layers of metals, both from a practical and theoretical point of view, have long been studied [1-5]. Mechanisms of growth of metal films were considered in detail in [6, 7, 8, 9].

An island growth mechanism is realized when the condition [10]:

$$\sigma_S < \sigma_d + \sigma_{S-d} - const \times k_B T ln(\xi + 1), \quad (1)$$

where σ_s - free energy per unit surface of the substrate, σ_d - free energy per unit surface film, $\sigma_{(s-d)}$ - the free energy per unit surface substrate-film section.

Experimental Section

Zn films were deposited on glass, quartz and silicon substrates measuring 20 × 20 mm^2 by magnetron sputtering at various currents and constant pressure Ar ~1 × 10^{-2} Torr. The source was a target of Zn with a diameter of 57 mm. The distance between the target and the substrates, which are fixed on a rotating holder holding at a constant speed, is ~5 cm. The deposition rate was measured. The film thickness was determined by an integrated meter, the value of which was taken as the "effective film thickness." In this case the effective thickness is equal to h_{ef} film: $h_{ef} = m/d \times S$, where m - mass of deposited metal, d - the density of the metal material, S - area occupied by the film on the measuring surface.

As can be seen from the SEM results in Figure 1 a, b and the histograms of the size distribution of islands of the film structure on different substrates differ greatly.

The average grain size on glass is 23.1 ±5.6 nm, on quartz 12.9 ± 0.6 nm. New atoms come from the vapor of the material during the sputtering of the target. Assuming that the nanocluster is a monolayer of Zn atoms, it consists of 300-1000 atoms [11].

With a further increase in the thickness of the Zn film, the size of the islands grows on a glass substrate. The morphology of the surface does not change significantly up to a thickness of 35 ± 40 nm.

The dependence of the film conductivity on the thickness is linear.

As the film thickness increases, more zinc nanoparticles begin to touch each other and an increasing number of zinc grains begin to participate in the conductivity of the film, which ultimately leads to a decrease in film resistance.

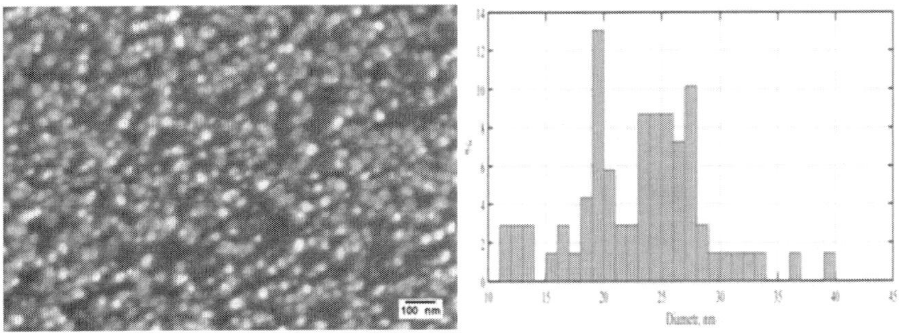

(a)

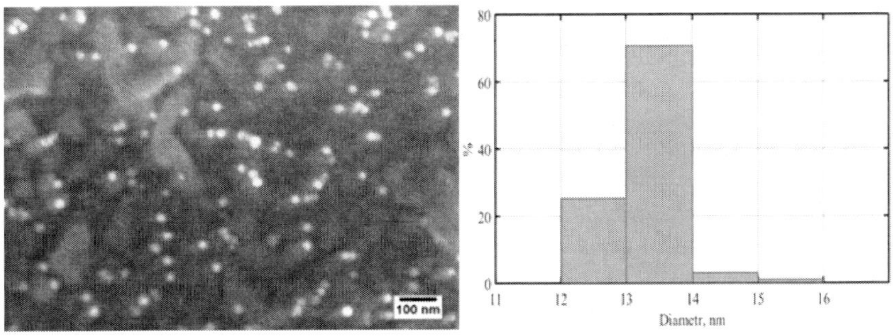

(b)

Figure 1. SEM view of the film structure on glass substrate (a) and quartz substrate (b).

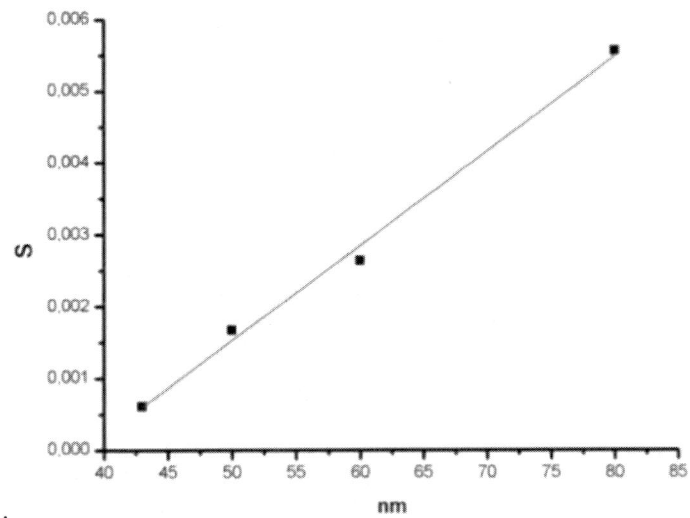

Figure 2. Dependence of the conductivity of Zn films on their thickness on quartz substrates.

Figure 2 shows the dependence of the conductivity of Zn films on their thickness on quartz substrates.

For thick zinc films (h > 100 nm), the conductivity of the film will tend to conduct the bulk polycrystalline zinc.

Conclusion

1. It is shown that on substrates of glass and silicon during magnetron sputtering, formed island films with an effective thickness of 10 ÷ 12 nm and an average grain size of 23.1 ± 5.6 nm on glass.
2. On the substrate of quartz, island films with an average size of 12.9 ± 0.6 nm islands are formed.
3. Oriented hexagonal Zn single crystals with the [0001] direction parallel to the plane of the substrate are formed.

Acknowledgment

This work was partly supported by Russian Science Foundation 17-03-00121.

References

[1] Levinstein H. 1949 *J. of Appl. Phys.* V 20 April 306.
[2] Konozenko I. D. 1954 *Physics-Uspekhi* V 52 N4 561.
[3] Yoo M. H. 1981 *Metallurgical Transactions A* V 12A N3 409.
[4] Kukuschkin S. A., Osipov A. V. 1998 *Physics-Uspekhi* V 168 N10 1083.
[5] Ohring M. 2002 *Material science of thin films Deposition and Structure*. Academic Press. USA.
[6] Volmer M., Weber A. 1926 Z. *Physik. Chem.* 119 277.
[7] Stranskiy I. N, Kaischev I. K. 1939 *Physics-Uspekhi* V 21 N4 408.
[8] Frank F. C. and van der Merwe J. H. 1949 One-dimensional dislocations. I. *Static Theory,* Proc. Roy. Soc. London, Ser. A 198 205.
[9] StranskiI. N., Krastanow L. 1938 *Abhandlungen der Mathematisch-Naturwissenschaftlichen Klasse IIb* [*Treatises of the mathematical-scientific class IIb.*]. Akademie der Wissenschaften Wien [Academy of Sciences Vienna]. 146 797.
[10] Venables J. A., Spiller G. D. T., Hanbücken M. 1984 *Rep. Prog. Phys.* 47 399.
[11] Ievlev V. M., Shvedov E. V. 2006 *Physics of the Solid State* V 48 N1 133.

In: Applied Aspects of Nano-Physics …
Editors: Kirill Levine and Andrey Syrkov
ISBN: 978-1-53614-708-7

Chapter 32

FEATURES OF METROLOGICAL SUPPORT IN THE IMPLEMENTATION OF NANOTECHNOLOGY: ACCOUNTING FOR NONLINEAR EFFECTS

Andrey Syrkov**, *A. A. Vinogradova and Kirill Levine
Saint Petersburg Mining University, St. Peterburg, Russia

ABSTRACT

The authors discussed the historical tradition of device development and control methods for nano-objects and the modern features concerning metrological support of the implementation of nanotechnology.

Keywords: nanotechnology, research methods, additives

INTRODUCTION

In recent years the leading manufacturers of scientific equipment was developed, high-precision instruments for the measurement and integrated interpretation of micro- and nanoscale mechanical and tribological properties of nanostructured materials and coatings based on the methods of measuring the indentation (penetration testing), scratching (adhesion test) and friction (tribological tests) were tested.

* Corresponding Author's Email: syrkovandrey@mail.ru.

DISCUSSION

For metrological measurements correct and commercialization of products of nanotechnology require a calibrated means of measurement, including standard samples intended for use in the system of ensuring the uniformity of measurements for verification, calibration, control of metrological characteristics of measuring instruments when conducting their tests, metrological certification of measurement procedures, monitoring the stability of the measurement results [1, 2, 3].

To date, only probe techniques using different nanoscopes, are unable to solve the problem of adequate control of surface-nanostructured metals and tribological materials based on them taking into account nonlinear effects. According to the author of the works [4, 5] to solve this problem, a promising approach is based on finding relationships between macroscopic properties, which are sensitive to changes in the composition of the surface layers of the substance.

For oil I-20 with additives dispersed metals containing on the surface of low-dimensional forms of ammonium compounds, the most sensitive proved to be an integral indicator of friction D measured in the range of acoustic frequencies, and its relationship with adsorption-chemical characteristics of dispersed additives [4-6].

In the Mining University this is more than a century long tradition of development of devices and control methods for nano-structured objects. It comes from P.P. Veimarn, its Professor and graduate. Some recognized experts in nanotechnology consider Veimarn as the founder of the science of nanotechnology, which is the first concept formulated basic principles of nanotechnology approach [7-9]. In the early twentieth century Weimarn actively used optical microscopy Sismondi and studied electrical conductivity of highly dispersed metals (1912) [10].

One of the main problems related to metrological assurance of nanotechnology is the problem of test object [2]. In the measurement of characteristics of nanoscale objects it is very difficult to observe the basic principle of Metrology: measuring instrument must have an accuracy high enough to provide valid measurement. The most difficult task is to find etalon (the transfer unit size) in the nanometer range. Practical implementation of finding etalon depend on the respective reference measurement [8, 9, 11]. According to [11], the most "prepared" method for solving the problem of control of the sizes of elements large-scale integrated circuits in microelectronics is the method of scanning electron microscopy; atomic force microscopy also is promising to solve the considered task.

Conclusion

Further developing an approach was described by Weimarn, with the problems arising in the measurement of nanometric objects being solved. Measuring D (the integral indicator of friction) in the range of acoustic frequencies was made possible, which is indicative to the structure of the monomolecular surface layer of the additive [12, 13].

Therefore, it is advised to use a friction method in the range of acoustic frequencies, as it was made to achieve full control of tribological parameters and make diagnostics of the tribological system with surface-modified additives [14].

References

[1] Blank, V. D. Gogolinsky K. V., Pyatov, A. L. *Legislative and applied Metrology.* 1 (107), (2010) 27-28.

[2] Kovalchuk M. V., Todua P. A. *Technosphere*, (2009) 136.

[3] Gogolinsky K. V. *Proceedings of the Mining Institute*, (2015) 264.

[4] Syrkov A. G. *Proceedings of the Mining Institute*, 216, 2015. 122-130.

[5] Syrkov A. G. Publishing house of Polytechnic University, (2016) 194.

[6] Nazarova E. A., Syrkov A. G., Brichkin V. N., *Advanced Materials Research*, 1040, (2014) 103–106.

[7] Zhabrev V. A., Publishing House "*Elmor*," (2012) 328.

[8] Margolin V. I. Publisher "*LAN*," (2012) 27.

[9] Syrkov A. G. *Condensed matter and the interface*, 19 (1), (2017) 140-147.

[10] Veimarn, P. P. Econ. *Tipo-Litor.,* 1912. 15.

[11] Novikov Y. A., Rakov A. V. *Measuring equipment*, 1, (1999) 14-18.

[12] Vinogradova A. A. Syrkov A. G., Simakov A. S. *Condensed matter and interphases,* 15 (2), (2013) 203-205.

[13] Vinogradova A. A., Boikov A. V., Syrkov A. G. *Competence,* 7, (2016) 12-16.

[14] Boikov A. V., Vinogradova A. A., Syrkov A. G. *Competence*, 9-10, (2016) 203-205.

In: Applied Aspects of Nano-Physics …
Editors: Kirill Levine and Andrey Syrkov
ISBN: 978-1-53614-708-7

Chapter 33

THE STUDY OF LOW-DIMENSIONAL SYSTEMS IN SAINT-PETERSBURG MINING UNIVERSITY: FROM P. P. WEIMARN TO THE PRESENT DAY

Andrey Syrkov[1,*] and I. V. Pleskunov[2]
[1]Mining University, Saint Petersburg, Russia
[2]IMC Montan, London, Great Britain

ABSTRACT

Major scientific laws, which were discovered by P. P. Weimarn (1879 - 1935) in the Saint-Petersburg Mining Institute in the field of physics and chemistry of disperse systems with nanoscale particles, were analyzed. Priority publications in the mentioned field were established (1906 - 1915). Features of P. P. Weimarn's scientific school and its connection with modern researches, which are conducted at the Mining University in nanotechnology were considered. The succession takes place in a number of objects (dispersed metals) of researches and in the methodology of studying properties of materials, depending on their dispersity. Information about the achievements of synthesis of nanostructured materials was given.

Keywords: scientific priority, low-dimensional systems, nanotechnology, P. P. Weimarn contribution, dispersoidology, surface and dispersity of substances, ultra-dispersed state

* Corresponding Author's Email: syrkovandrey@spmi.ru.

INTRODUCTION

The question of priority in the field of science of nanotechnology is very important and can be widely discussed. However, this question is interesting not only for the history of science and not only for the scientists, but for the numerous people in the case of determination of scientific priority between researchers, institutes and countries in laying down the foundations of the "nanotechnologies and nanomaterials" direction. Such outstanding experts as V. A. Zhabrev, V. T. Kalinnikov, V. I. Margolin, etc. published several researches, which declared that more than a hundred years ago the professor of the Saint-Petersburg Mining Institute Peter Weimarn (Figure) postulated the basics of the nanotechnology as a science [1, 2].

The aim of this research is based on the answering of important questions that were not discussed in detail in scientific literature:

- What had Peter Weimarn done exactly in the field of nanotechnology? What laws had he discovered? Where had he firstly published his major researches?
- Who was his scientific head? Which scientific school had he belonged to? Had he founded his own scientific school?
- What was the impact of his researches on the current works in the field of nanotechnology and especially in the works of modern researchers from the Saint-Petersburg Mining University? What are their prospects and orientation?

Figure 1. P. P.Weimarn (1879 - 1935).

RESULTS AND DISCUSSION

Before 2012, many researchers of P. P. Weimarn's scientific papers mistakenly considered him only as a founder of colloid chemistry, and thought that his achievements related only to this field of science [3, 4]. In fact, T. Graham founded the colloid chemistry in 1862. Actually, Weimarn's scientific interests laid far beyond colloid chemistry. He was the founder of new field of science – "dispersoidology", which studies properties of surface and processes, which take place on it [5]. Field of science, which Weimarn called "dispersoidology", modern researchers would name physics and chemistry of surface of disperse substances. It is obvious that mentioned field of science is a fundamental basis of nanotechnology.

Major Weimarn's achievements during the period of his work in Saint-Petersburg Mining Institute (1908 - 1915) are listed below:

- He postulated that between the world of molecules and microscopically visible particles exist a special form of matter with its genuine complex of new physical and chemical properties. It was named ultra-disperse or colloidal state of matter, which appears if dispersity lies in the range between 10^{-5} and 10^{-9}m. In this state of matter, the characterizing size of films (thickness), fiber and particles ranges from 1 to 1000nm.
- He formulated the basic law of dispersoidology: in the process of physico-chemical or mechanical grinding of substance, it tends to turn into modifications or states with the lowest surface energy; these modifications and states have the lowest surface tension and, in most cases, the lowest density as well [5].
- During the study of more than 200 disperse systems, it was revealed that there are special conditions under which the process of dissolution or crystallization of the substance accompanied by the formation of highly-disperse stable systems (particle size is less than 0.1 microns) [6, 7].
- It was found that with increase of dispersity of chemically pure metals, the electrical resistance increases and the temperature coefficient decreases.
- The basic ideas and principles of operation of the electric ultramicroscope, which allows to explore the opaque material (metals) were formulated [8].
- The principles and methods of modern nanotechnology approach were conceptually formulated [1, 2, 9].

In the mentioned period, P. P. Weimarn published 26 articles in scientific journal "Notes of Mining Institute". It is important to note that 17 of them described the problems of dispersoidology [9]. These priority articles contained the confirmation of the achievements listed above. Almost at the same time P. P. Weimarn had sent his papers to be published in journal "Kolloid-Zeitschrift" (Germany), which published 211 his articles in the period from 1907 to 1935. From 1908 to 1912 six Weimarn's scientific monographs were published in Germany, including publishing house "Teodor Steinkopff". At that time, German was the language of scientific communications. Therefore, from 1916 to 1920, Weimarn's scientific researches received world recognition and gained fame abroad [1-3, 9]. The Nobel Prize winner - V. F. Ostvald thought that Weimarn was the genius scientist. Australian mineralogist Felix Corn named the first colloid mineral "Weimarnit" in his honor. Japanese professor K. Kashima noted that researches which were conducted by Weimarn in Japan, after his immigration in 1921, characterized him as an outstanding chemist [3]. Some of results of his researches were successfully used in the industry of Japan. In English version of Wikipedia presented a von Weimarn law (1906): *Sols are obtained from very dilute or very concentrated solutions but not from intermediate solutions. The relative supersaturation ratio herein is defined by* $S = (Q-L)/L$ *(where Q is the amount of the dissolved material and L is its solubility).* The von Weimarn law is still relevant as the basis of sol-gel technologies, which are one of the methods of modern nanotechnology.

Weimarn's teachers were academician N. S Kurnakov and professor I. F. Schreder, who was a director of Mining Institute from 1912 to 1918. Being a student, Weimarn had written and published his first articles in collaboration with Kurnakov in 1902. Weimarn rarely published articles in collaboration with his own scholar. However, his assistant I. B. Kagan was an exception [7]. Generally, Weimarn's disciples (I. A. Avalov, S. Ya. Levites, N. I. Morozov, K. D. Lugovkin, B. V. Byzov A. M. Yanek, etc.) helped him in the calculations or conducted minor experiments. Weimarn used to conduct major experiments and write articles by himself. Only after practice, he allowed his student to work by themselves.

One of the most talented disciples, A. M. Yanek published a book called "Dispersoidology" (1915) in Saint-Petersburg by himself. N. V. Hisamutdinova wrote with reference to Weimarn's colleagues that "He created his scientific school, which consisted of talented students who successfully developed the ideas of their teacher" [4]. Despite the revolution and the civil war (1917 - 1921), Weimarn's disciples continued developing his ideas. Some of them (K. D. Lukovkin, N. I. Morozov, A. M. Yanek) moved with him to Ural Mining Institute (Yekaterinburg, 1917 - 1919) and then to Vladivostok Polytechnic Institute (1919 - 1921). Therefore, it is obvious that Weimarn founded his own scientific school, due to the facts listed below: the presence of intergenerational continuity, international recognition of the results of his researches, general scientific activity, stability of scientific reputation, reproduction of highly qualified researchers, also because Weimarn followed the best traditions of scientific school of physico-chemists of Saint-Petersburg Mining University (Hess, Kurnakov, Schreder). The object (metals) of this scientific school and methods, which were used by Weimarn, are still being used by modern researches in Saint-Petersburg Mining Institute

(University since 2012) [9]. Particularly, researchers from the university (direct followers of Aleskovsky's scientific school) are studying the disperse metals and the relations between the properties of solids and their dispersity [9, 10]. Noted scientific and methodological features are common for the number of Weimarn's publications [5-8].

Currently, the research of low-dimensional systems in Saint Petersburg Mining University are being carried out in the following areas:

- Nanomineralogy – the study of nanostructures of minerals by using modern physical methods
- Nanometallurgy - development of new methods of synthesis of nanostructured composite materials (ligatures) based on light and rare earth elements
- Application the concepts of synergetic and fractal analysis to describe nanospace of coals
- Nanostructured regulation of antifriction and protective properties of the surface of the metal and its study by using special precision methods (XPS, AFM, EDX)
- Establishment and characterization of new nanostructured metals (Ni, Fe, Cu, Al) with protective nanofilms
- Development of scientific foundations of electroadhesion technology of joining of different materials
- Plasma nanotechnology
- Creating nanostructured information fields on the surface and in the volume of crystalline materials
- Nanotechnology to produce a new generation of electronic systems
- The mechanism of intercalation of lithium into carbon-graphite lining of metallurgical units
- Methods of control the geometrical and mechanical properties of solids with micro-and nanometer resolution

According to the traditions of Mining Institutes, in modern Mining University a number of pioneer researches in the field of nanotechnology were carried out and methods of obtaining of low-dimensional materials were developed:

- layering different-sized molecules of ammonium and silicon-organic compounds on metals [10, 11];
- solid state hydride synthesis of surface nanostructured disperse metals [12].

Moreover, original areas of research, which were developed, are listed below:

- the influence of reducing agent on structure and reactivity of disperse metals;

- nonlinearity of properties of surface-modified metals in nonequilibrium systems [13, 14].

CONCLUSION

1. Professor of Saint-Petersburg Mining Institute, P. P. Weimarn – the founder of nanotechmology, more than a hundred years ago (1906 - 1915) postulated laws, conditions and the method of obtaining a colloid solution with adjustable dispersity of solid phase (up to molecular) [6].
2. During the process of studying of electrical resistivity of disperse metals, P. P. Weimarn proposed the ideas and basics of electrical ultramicroscope with accuracy of measurements of size of the particles up to nanometers.
3. His first articles about dispersoidology and obtaining nanoscale particles were published in "Notes of Mining Institute".
4. His first scientific research in collaboration and under supervision of N. S. Kurnakov was published in Journal of Russian physico-chemical society [9]. In addition, one of Weimarn's preceptors was Professor I. F. Schreder.
5. P. P. Weimarn continued and developed traditions of well-known the scientific school of the physico-chemists of the Mining Institute under the guidance of academic N. S. Kurnakov. Weimarn succeeded in establishing his own scientific school in the field of dispersoidology (physics and chemistry of surface of disperse matter), which are the basics of modern nanotechnology. The interesting feature of this scientific school in that in fact it was international, because P. P. Weimarn actively worked with researches from German universities (Wo. Ostvald and others) and with Japanese post-graduate students (1921 - 1935) [4, 9].
6. There is the succession between modern researchers from the Mining University and Weimarn's scientific works in objects and methods of studying. Among the researches of the 21st century the most perspective are: Solid state hydride synthesis of surface-nanostructured disperse metals and molecular layering of different-sized molecules on metals (Al, Cu, Ni, Fe).

ACKNOWLEDGMENTS

The work was carried out with the financial support of the Ministry of Education and Science of the Russian Federation (State contract № 14.577.21.0127 from 20 October 2014. Unique identifier of applied research RFMEFI57714X0127).

REFERENCES

[1] V. A. Zhabrev, V. T. Kalinnikov, V. I. Margolin, A. I. Nikolaev, V. A. Tupik. *Physico-chemical processes of the synthesis of nanoscale objects*. St. Petersburg: Publishing House "Elmore". 2012. p. 328.

[2] V. I. Margolin, V. A. Zhabrev, G. N. Lukyanov, V. A. Tupik. *Introduction to nanotechnology*. SPb: Publishing House "Lan", 2012. p. 464.

[3] K. Kashima. An Eminent Chemist. *Industrial and Engineering Chemistry*, 1924. V.16. p. 540.

[4] N. V. Hisamutdinova. Chemist Petr Petrovich von Weimarn in Russia and Japan. *Bulletin of the Far Eastern Branch of the Russian Academy of Sciences,* 2011. No. 5. p. 134 - 141.

[5] P. P. Weimarn. The new classification of aggregate states of matter and the fundamental law dispersoidology. *Notes of Mining Institute*, 1912 V.4. (2). p. 128 - 143.

[6] P. P. Weimarn, I. B. Kagan. A simple common method of obtaining any object in any state of solid colloidal solutions of any dispersity ranging from molecular. *Notes of Mining Institute*, 1910. V.2. (5). p. 398 - 400.

[7] P. P. Weimarn, I. B. Kagan Dispersoid chemistry of copper chloride in benzene. *Notes of Mining Institute*, 1912. V.4. (2). p.75 - 95.

[8] P. P. Weimarn. About the electrical conductivity of metals and their alloys from dispersoid chemistry point of view. *Notes of Mining Institute*, 1911. V.3. p. 349 - 353.

[9] A. G. Syrkov. About priority of Saint Petersburg Mining University in the field of nanotechnology science and nanomaterials. *Journal of Mining University,* 2016, V.221. p.730 - 736.

[10] A. G. Syrkov. Synergetic improvement of aluminium reactivity in the presence on the surface of quaternary ammonium compounds. *Russian Journal of General Chemistry*, 2013, Vol.83, No 8, p. 1621.

[11] A. G. Syrkov. Synergetic change of tribochemical properties of copper in the presence of quaternary ammonium compounds at the surface. *Russian Journal of General Chemistry*, 2015. Vol. 85, No. 6, p. 1536.

[12] A. G. Syrkov. Surface-nanostructured metals and their Tribochemical Properties (book Chapter) *Smart nanoobjects: from laboratory to industry*. New York; Nova Science Publishers, Inc. 2013. p. 214.

[13] E. A. Nazarova, A. G. Syrkov, V. N. Brichkin. Nonlinearity of dependence of Integral index of friction of tribosystem from Hydrophilic Properties of Surface-Modified metal Fillers. *Advanced Materials Research*, 2014. V.1040. p. 103.

[14] A. G. Syrkov, M. O. Silivanov, A. N. Kushenko. Tribochemical peculiarities of lubricant composition with surface-modified metal powder. *Journal Physics Conf. Ser.*, 2016. V.729. p. 012026.

In: Applied Aspects of Nano-Physics ...
Editors: Kirill Levine and Andrey Syrkov
ISBN: 978-1-53614-708-7

Chapter 34

LOW-FREQUENCY DIELECTRIC SPECTROSCOPY OF THE $(As_2Se_3)_{100-x}Bi_x$ GLASSY SYSTEM

Rene A. Castro**, Gennady A. Bordovsky and Nadezhda I. Anisimova
Herzen State Pedagogical University of Russia,
Saint Petersburg, Russia

ABSTRACT

Results of research into the processes of dielectric relaxation in a glassy system $(As_2Se_3)_{100-x}Bi_x$ are presented. In the low-frequency region, the presence of dielectric maximum was identified, which was determined by the existence of defect charged centers Bi_3^+ and/or Bi_4^+. At high concentrations of Bi impurity, there appear micro-heterogeneous regions (clusters) composed of Bi_2Se_3. The size of clusters is d = 3 ... 5 nm in the Debye approximation.

Keywords: low-frequency, dielectric spectroscopy, chalcogenide glassy semiconductors, defect and impurity centers

INTRODUCTION

Chalcogenide glassy semiconductors (CGSs) attract the attention of researchers due to their application in multiple micro- and optoelectronic devices. They can be used as

* Corresponding Author's Email: recastro@mail.ru

source materials in manufacturing fiber-optic light guides and infrared fiber-optic lasers [1-4]. These compounds have good potential for constructing solar cell elements [5].

An overview of early studies on the influence of impurities introduced by the method of synthesis on the electrical conductivity of CGSs was presented in the paper [6], along with the communication that these materials could not be doped. The Mott-Gubanov theory [7, 8] explained this statement by the fact that under conditions of the disordered CGS structure, the impurity atom has the possibility to use all its valence electrons to form bonds with neighbors. This statement was later formulated as the "rule of 8-N" [9] and asserted that an atom in a glassy state, having N valence electrons (when $N \geq 4$), always forms valence bonds with $8 - N$ neighbors. Therefore, impurity atoms in the disordered glassy network are capable of filling all valence $(8 - N)$ bonds and, in this case, become electrically inactive.

However, it turned out later that the situation is reversed if a non-equilibrium method of sample preparation is applied. A radiofrequency (high-frequency) co-dispersion of CGS and impurity onto the cold substrate (the so-called modification method) made possible producing impurity conductivity [10]. Moreover, the conductivity type could be changed by increasing the concentration of the introduced impurity. These challenges of getting new semiconductor materials characterized by p- and n-type of conductivity, which in turn creates the prerequisites for constructing CGS-based p-n transitions, gave a fresh impetus to study these structures. To date, the electrical conductivity in doped CGSs, mobility of charge carriers, electron- and photo-optic properties, etc., are considered in detail. However, there is still lack of studies on polarization of these materials.

The objective of this study was to identify specific features of dielectric relaxation processes in thin layers of $(As_2Se_3)_{100-x}Bi_x$ glassy system and their relationships with structural features of the studied systems.

EXPERIMENTAL SECTION

Capacitor structures Al-$(As_2Se_3)_{100-x}Bi_x$-Al ($x = 0.5, 1.5, 2.5$ at. %) used for dielectrical measurements were prepared by RF sputtering at UMR-3-021 ionic-plasma RF sputtering unit. The thickness of $(As_2Se_3)_{100-x}Bi_x$ layers was measured with the ELF spectroellipsometer and made ~ 1.0 μm and the overlapping area of electrodes was ~ 14.0 mm^2 [11, 12]. Temperature-frequency dependences of the complex dielectric permittivity components were measured in the studied layers with "Concept-81" spectrometer (Novocontrol Technologies GmbH) (Figure 1) designed for studying the dielectric and electrical properties of an extensive class of materials. Measurements were taken at wide ranges of frequency $f = 10^{-2}...10^{6}$ Hz and temperature $T = 243... 353$K. The voltage U =

10^{-1} V was applied to the samples. Physical measurement errors and calculated parameter errors did not exceed 3%.

The spectra of complex dielectric permittivity and complex conductivity were calculated from the impedance spectra:

$$Z^*(\omega) = R + \frac{1}{i\omega C} = Z' + iZ'' = \frac{U_0}{I^*(\omega)} \tag{1}$$

according to the formulas:

$$\varepsilon^* = \varepsilon' - i\varepsilon'' = \frac{-i}{\omega Z^*(\omega)} \frac{1}{C_0} \tag{2}$$

$$\sigma^* = \sigma' - i\sigma'' = \frac{-i}{\omega Z^*(\omega)} \frac{S}{d} \tag{3}$$

where $C_0 = \varepsilon_0 S / d$ is the capacity of the empty cell.

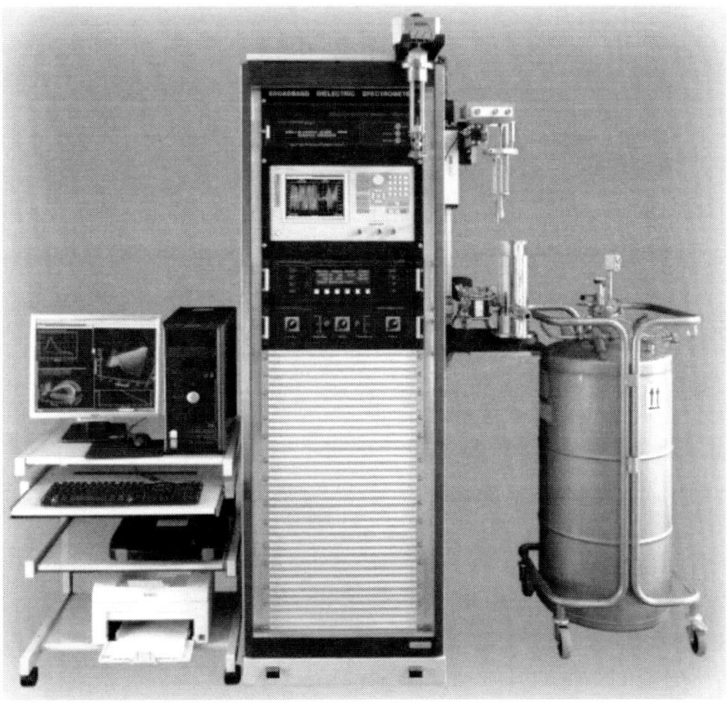

Figure 1. Broadband dielectric spectrometer "Concept-81" (Novocontrol Technologies GmbH).

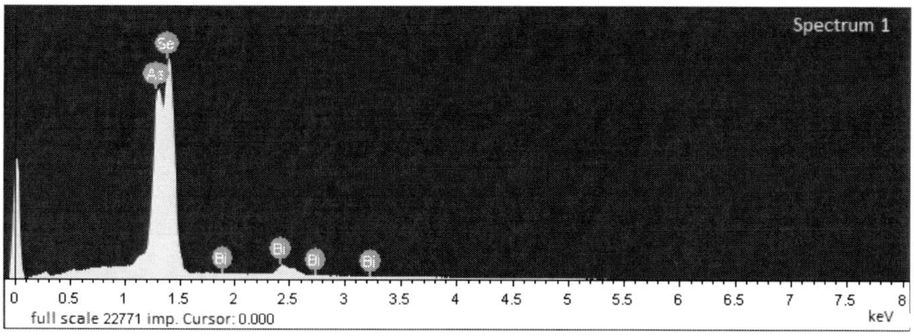

Figure 2. Spectrum of the atomic content of thin layers in (As2Se3)100-xBix glassy system (x = 2.5 at. %).

Table 1. Elemental composition of the samples of the system $(As_2Se_3)_{100-x}Bi_x$ (x = 2.5 at.%)

Element	Nominal concentration	Mass fraction	Weight %	Weight % Sigma	Atomic fraction %
As L	45.70	1.7588	33.09	0.26	36.96
Se L	31.62	0.6906	58.32	0.32	60.47
Bi M	4.75	0.7034	8.60	0.33	2.57
Total			100.00		

The elemental composition of samples was studied with Carl Zeiss EVO 40 scanning electron microscope (SEM) (Figure 2, Table 1).

RESULTS AND DISCUSSION

The temperature dependence of dielectric losses of $(As_2Se_3)_{100-x}Bi_x$ samples reveals existence of relaxation process in the field of low frequencies (Figure 3). Increase in percentage of bismuth leads to the shift of a maximum of losses to the range of more high temperatures.

Frequency dependence of the dielectric loss tangent in $(As_2Se_3)_{100-x}Bi_x$ (x = 0; 1.5at. %) samples is characterized by the presence of a single low-frequency maximum (Figure 4, 5), which is most probably associated with the manifestation of the dipole-relaxation polarization mechanism. Relaxation centers can be represented by both defects of the inherent glass structure and those associated with doping the glass matrix.

In many disordered systems, the permittivity dispersion and the existence of the loss maximum are supposed to be associated with the presence of a complex spectrum of relaxators in the structure. To study the distribution pattern of relaxators in thin layers of $(As_2Se_3)_{100-x}Bi_x$ glassy system, with reference to the relaxation times, we applied a two-parametric Havriliak-Negami (HN) function [13]

$$\varepsilon^*(\omega) = \varepsilon_\infty + \frac{\Delta\varepsilon}{\left[1 + (i\omega\tau)^{\alpha_{HN}}\right]^{\beta_{HN}}} \tag{4}$$

where, ε_∞ is the high-frequency limit of the real part of dielectric permittivity, $\Delta\varepsilon$ is the dielectric increment (difference between the low- and high-frequency limits), $\omega = 2\pi f$, α_{HN} and β_{HN} are shape parameters that describe the symmetric ($\beta = 1.00$, the Cole-Cole distribution) and asymmetric ($\alpha = 1.00$, the Cole-Davidson distribution) expansion of the relaxation function, respectively. The values of relaxation parameters for samples with varying iron percentage are presented in Table 2. Approximation of the experimental curves by the HN function lead us to the conclusion that the system under study demonstrates a non-Debye oscillatory process, where the relaxation times are distributed according to the Cole-Cole model for the case of symmetric distribution relaxators.

To determine the activation energy of relaxation processes in $(As_2Se_3)_{100\text{-}x}Bi_x$ system, we examined the temperature dependence of the most probable relaxation time τ_{max}, determined under the HN approximation (Figure 6). Dependences $-\log\tau_{max} = \varphi(1/T)$ are linear and can be described by the Arrhenius equation:

$$\tau(T)_{max} = \tau_0 \exp\left(\frac{E_a}{RT}\right) \tag{5}$$

where, $\tau_0 = \tau_{max}$ at $T \to \infty$, E_a is the activation energy of the relaxation process. The equation parameters (4) are presented in Table 2.

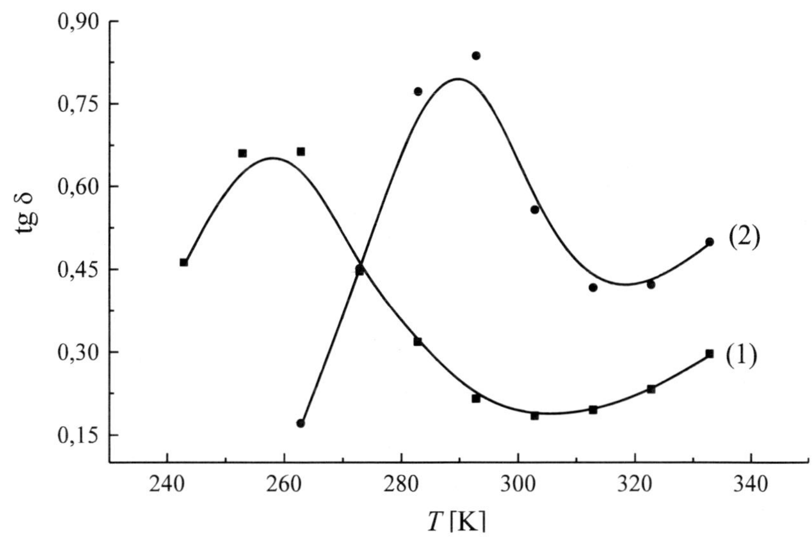

Figure 3. Temperature dependence of dielectric loss for (As2Se3)100-xBix samples at frequency f = 10-2 Гц. (1) – x=1.5 at. %, (2) – x = 2.5 at. %.

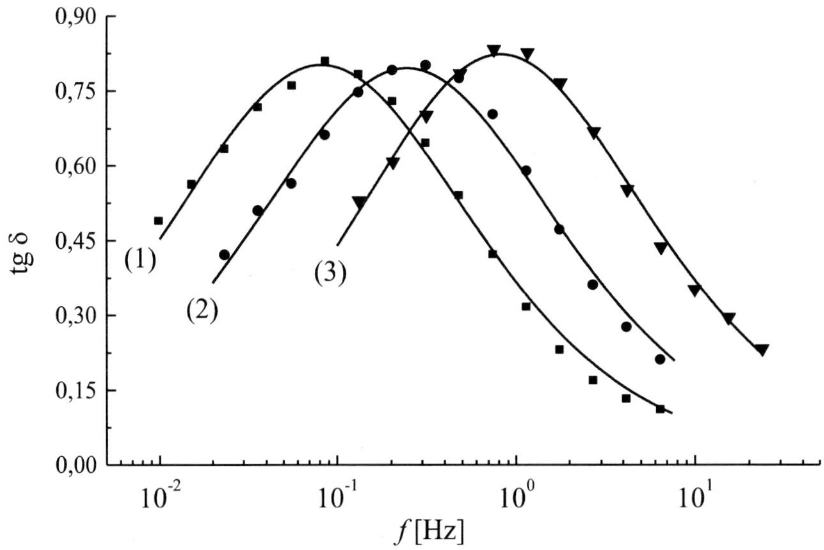

Figure 4. Frequency dependence of dielectric loss for (As2Se3)100-xBix (x=0.5 at. %) samples at different temperatures: (1) – T = 313 K, (2) – T = 323 K, (3) – T = 333 K. Solid lines - approximation by the HN function.

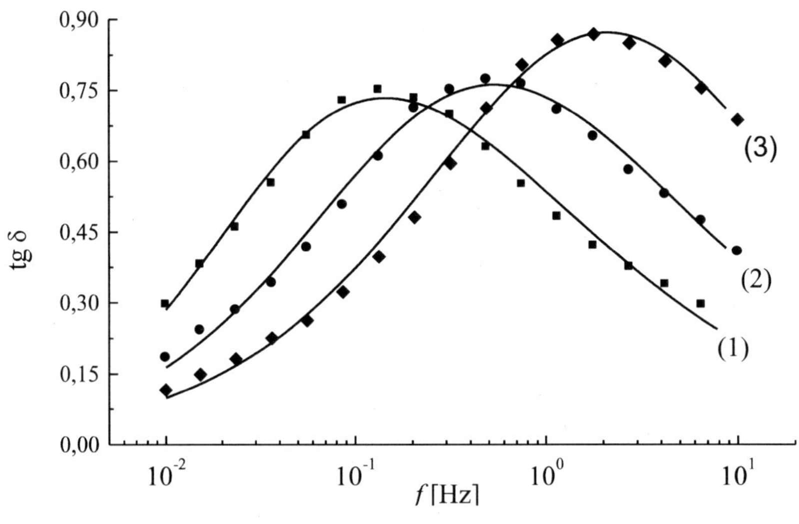

Figure 5. Frequency dependence of dielectric loss for (As2Se3)100-xBix (x=2.5 at. %) samples at different temperatures: (1) – T = 283 K, (2) – T = 293 K, (3) – T = 303 K. Solid lines - approximation by the HN function.

Introducing bismuth impurity into the glassy network results in the linear decay of the activation energy E_a of the relaxation process, frequency of loss maximum f_m grows with increase in percentage of impurity (Figure 7, Table 2). The sensitivity of mentioned parameters to varying impurity content can be used for control of amount of impurity in the experiments on studying of electronic properties of the studied binary chalcogenide system.

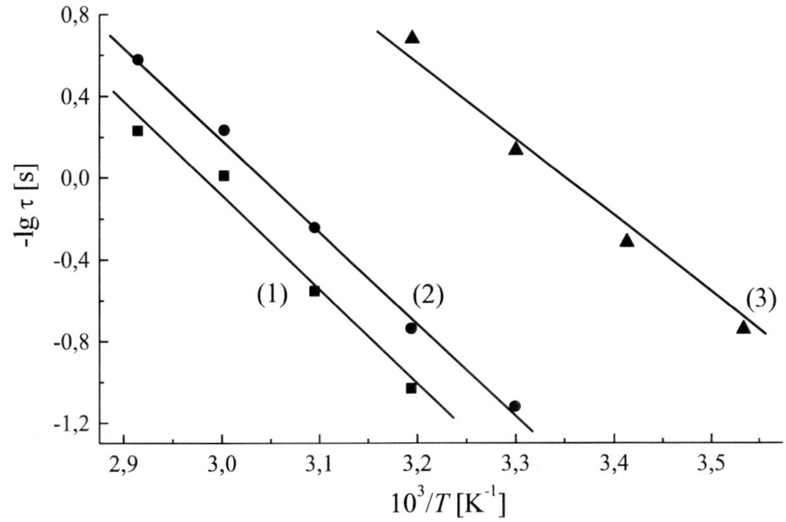

Figure 6. Temperature dependence of the most probable relaxation time τmax for (As2Se3)100-xBix samples: (1) – x = 0.5 at. %, (2) – x = 1.5 at. %, (3) – x = 2.5 at. %.

Table 2. Value of relaxation parameters of $(As_2Se_3)_{100-x}Bi_x$ samples at temperature T = 313K

Bi content (at. %)	$_{max}$ (c)	Δε	α_{HN}	β_{HN}	E_a (эВ)	N(E) [$cm^{-3}eV^{-1}$]
0.5	10.88	9.43 ± 0.39	0.72 ± 0.10	1.00 ± 0.15	0.93 ± 0.06	$6.2*10^{17}$
1.5	5.53	10.67 ± 0.78	0.75 ± 0.07	1.00 ± 0.16	0.90 ± 0.03	$3.1*10^{19}$
2.5	0.21	18.20 ± 1.25	0.76 ± 0.05	1.00 ± 0.14	0.83 ± 0.05	$2.5*10^{22}$

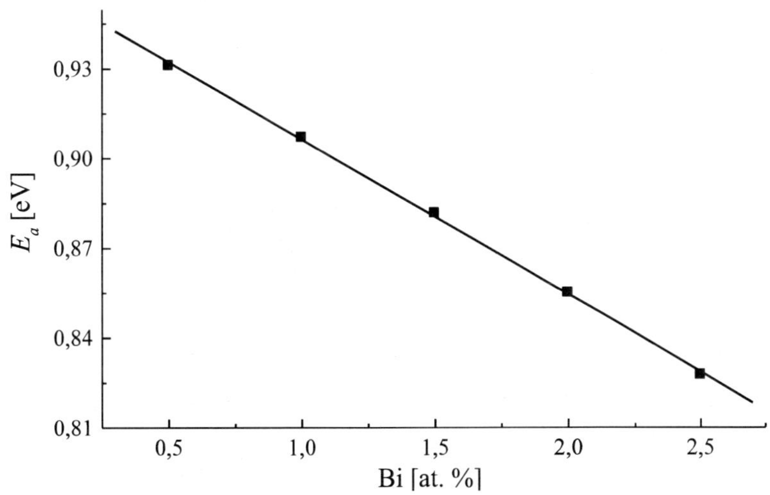

Figure 7. Dependence of activation energy of relaxation process on bismuth percentage in (As2Se3)100-xBix samples.

As it was noted earlier, relaxation centers can be represented by both defects of the inherent glass structure and those associated with the introduction of an impurity into the glass matrix. Some authors suggest that low concentrations of bismuth, when penetrating into the original matrix of the amorphous arsenic triselenide, generate charged centers of Bi_3^+ and/or Bi_4^+ (the latter as with As_4^+) [14]. At the same time, the internal arrangement of the original composition does not undergo major changes.

At higher concentrations of Bi (x = 2.5 at. %), additive, we observed the appearance of the second maximum on the tanδ (f) dependence curve. It is known that, when atoms of an additive are injected into CGSs, perturbations of the disordered structure get increased around them causing the emergence of large-scale changes in the potential [15]. In this case, micro-heterogeneous regions with higher additive concentration can appear in the baseline glass matrix. These crystalline regions (clusters) composed of Bi_2Se_3 have a lower band-gap [16] and are characterized by higher coordination of atoms, as compared to the matrix. It can be assumed that specifically these regions are responsible for the changing pattern of dielectric relaxation processes and the appearance of the second maximum on the tanδ (f) dependence curve.

From the value of dielectric losses it is possible to estimate the quantitative changes in the spectrum of the localized states and size of the clusters [16]:

$$tg\delta = \left(\frac{1}{\varepsilon}\right)[N(E_f)]^2 kTe^2a^5\left[\ln\left(\frac{\nu_p}{\omega}\right)\right]^4, \tag{6}$$

where a – parameter, ν_p – phonon frequency, $N(E_f)$ – density of localized states at Fermi level (Table 2). An estimate of Bi_2Se_3 cluster sizes given in the Debye approximation provides the size of the crystalline region radius d in the interval of 3-5 nm.

The existence of a crystalline phase in the form of clusters is confirmed by the analysis of the temperature dependence of specific conductivity in the frequency range corresponding to the position of the second maximum. An exponential dependence of conductivity is observed with the changing slope of $\sigma' = f(10^3/T)$ curve at the temperature of phase transition in Bi_2Se_3 crystals (Figure 8). According to the authors [17], at temperatures below 300K, metallic conductivity appears in Bi_2Se_3 crystals and heating above this temperature initiates the semiconductor section of σ′ temperature dependence. The metallic character arises in Bi_2Se_3 due to the half filled 6p band overlapping the filled 6s band. It is also stated the presence of interstitial defects may result in ionized states with rise in temperature and provide free electrons for the conduction mechanism, thereby decreasing the absolute value of the electrical resistivity.

Existence in a glass matrix of high-conductivity metal inclusions is confirmed also by emergence of a lineal section on dependence ε' = f(σ') (Figure 9). It is possible to assume that on border of the areas which are characterized by the increased coordination of atoms with the main matrix, there are potential barriers which also make impact on process of exchange of charge carriers between the localized centers, and in general on polarizability of the studied systems.

The approach described above can be used for studying structural features of complex glassy systems [18], when additives of diverse nature are introduced, as well as in studies of their crystalline phases.

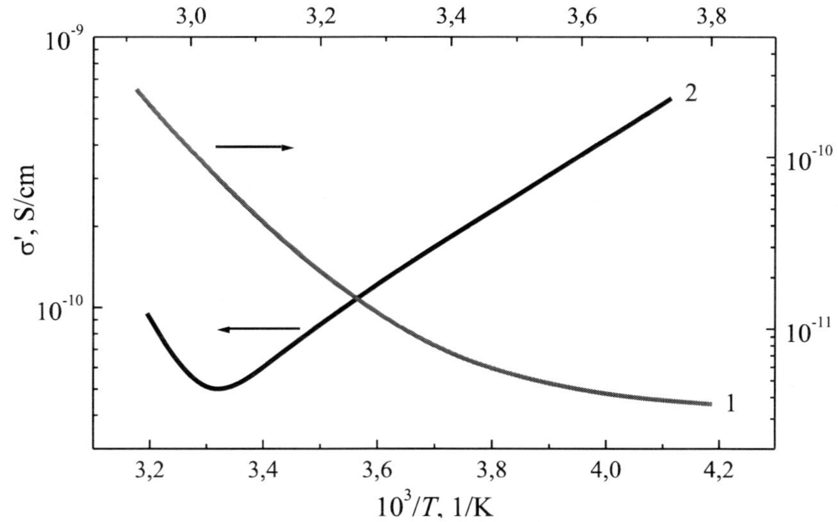

Figure 8. Temperature dependence of specific conductivity σ′ in (As2Se3)100-xBix samples: 1 – x = 1.5 at. %, 2 – x = 2.5 at. %.

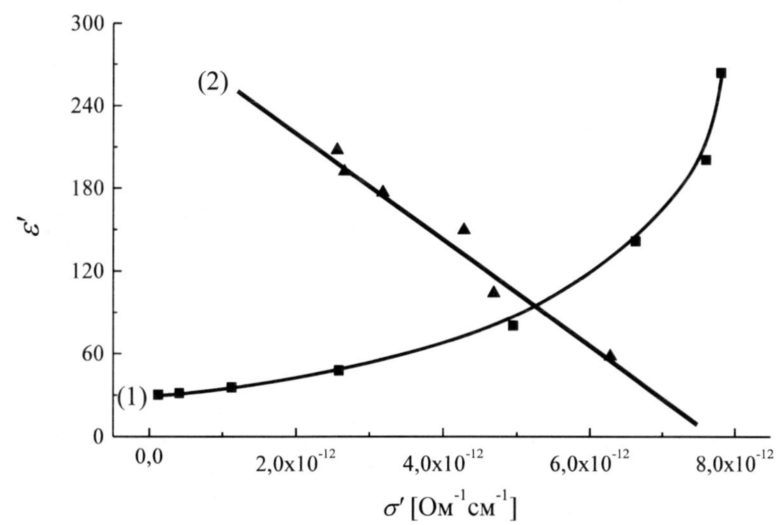

Figure 9. Dependence ε' = f(σ') for (As2Se3)100-xBix samples: 1 – x = 1.5 at. %, 2 – x = 2.5 at. %.

CONCLUSION

Thus, research into the processes of dielectric relaxation in $(As_2Se_3)_{100-x}Bi_x$ glassy system demonstrated the existence of dielectric loss maxima, indicating the existence of relaxation process in the low-frequency region. The tanδ maximum at x = 1.5 at. % bismuth is a manifestation of mechanisms of the dipole-relaxation polarization. Relaxation centers can be represented by both defects of the inherent glass structure and those associated with the introduction of bismuth additive into the glass matrix (e.g., Bi_3^+ and/or Bi_4^+ charged centers). At higher concentrations of Bi additive (x = 2.5 at. %), crystalline regions (clusters) of Bi_2Se_3 composition appear in the structure; those are responsible for changing pattern of the dielectric relaxation processes and the appearance of the second maximum on the tanδ (f) dependence curve. An estimate of Bi_2Se_3 cluster sizes given in the Debye approximation provides the size of the crystalline region within the value range from 3 to 5 nm. The described method can be used for studying structural features of complex glassy systems, when additives of diverse nature are introduced.

ACKNOWLEDGMENTS

The reported study was supported by the Ministry of Education and Science of the Russian Federation (project № 3.5005.2017/BY).

REFERENCES

[1] Kolobov, A. V., Fons, P., Frenkel, A. I., Ankudinov, A. L., Tominaga, J., Uruga, T. *Nature Materials*. 2004, vol. 3, pp. 703-708.

[2] Musgraves, J. D., Carlie, N., Guery, G. Chalcogenide glasses and their photosensitivity: Engineered materials for device applications, Bragg Gratings, Photosensitivity, and Poling in Glass Waveguides. OSA Technical Digest (CD). *Optical Society of America*, BWD1, 2010.

[3] Devaux, E., Vaux, E., Laluet, J. Y., Stein, B. *Optics express*. 2010, vol. 18(20), pp. 20610-20619.

[4] Šiljegović, M. V., Petrović, S. L., Petrović, D. M., Sekulić, D. L., Štrbac, G. R., Skuban, F. *Journal of Non-Crystalline Solids*. 2017, vol. 457, pp. 152-156.

[5] Kumar, S., Mehta, B. R., Kashyap, S. C., Chopra, K. L. *Applied physics letters*. 1988, vol. 52(1), pp. 24-26.

[6] Kolomiets B. T. *Physica Status Solidi. 1964*. vol. 7(2), pp. 713-731.

[7] Gubanov A. I. *Technical Physics*, 1957, vol. 27, pp. 2510.

[8] Mott N. F. *Advances in Physics*. 1967, vol. 16, pp. 49-51.

[9] Mott H., *Electrons in disordered structures*, MIR: Moscow, 1970.

[10] Fritzsche H., Kastner M. *Philosophical Magazine* B. 1978, vol. 37(2), pp. 127-133.

[11] Avanesyan, V. T., Bordovskii, G. A., Castro, R. A. *Glass Physics and Chemistry*. 2000, vol. 26(3), pp. 257-259.

[12] Castro, R. A., Bordovsky, G. A., Bordovsky, V. A., Anisimova, N. I. *Journal of non-crystalline solids*. 2006, vol. 352(9-20), pp. 1560-1562.

[13] Kremer, K., Schonhals, A., *Broadband Dielectric Spectroscopy*, Springer: Berlin, 2003.

[14] Saiter, J. M., Derrey, T., Vautier, C. *Journal of Non-Crystalline Solids*. 1985, vol. 77, pp. 1169-1172.

[15] Tsendin, K. D. *The electronic phenomena in the chalcogenide vitreous semiconductors*, Nauka: Saint-Petersburg, 1996.

[16] Austin, I. G., Mott, N. F. *Advances in physics*. 1969, vol. 18(71), pp. 41-102.

[17] Deshpande, M. P., Pandya, N. N., Parmar, M. N. *Turkish Journal of Physics*. 2009, vol. 33(3), pp. 139-148.

[18] Bordovskii, G. A., Castro, R. A., Marchenko, A. V., Seregin, P. P. *Glass Physics and Chemistry*. 2007, vol. 33(5), pp. 467-470.

In: Applied Aspects of Nano-Physics …
Editors: Kirill Levine and Andrey Syrkov

ISBN: 978-1-53614-708-7

Chapter 35

DIELECTRIC RELAXATION PROCESSES IN A GE28.5PB15S56.5 GLASSY SYSTEM

Rene A. Castro*
Herzen State Pedagogical University of Russia, Saint Petersburg, Russia

ABSTRACT

Low-frequency processes of dielectric relaxation were studied in Ge28.5Pb15S56.5 glassy system. The existence of distribution of relaxators over relaxation times is found, in line with the Cole-Cole model for the case of symmetric distribution of relaxation times. The activation energy of dielectric relaxation was equal to Ep = (0.40 ± 0.01) eV. The observed patterns are explained in terms of the model, according to which the structure of chalcogenide glasses represents a set of dipoles formed by charged defects of D+ and D- types.

Keywords: dielectric relaxation, glassy system, low frequency process, thin films

INTRODUCTION

Chalcogenide glassy semiconductors (CGS) of complex composition attract attention of researchers due to their wide application in micro- and optoelectronic devices [1-3]. The processes of charge transfer and accumulation in CGS of various systems can be associated with the exchange of electrons between the charged defects, which can be represented by defects of types D^+ and D^- being the centers with negative correlation energy [4]. The existence of defect states with activation energy of 0.43 eV and

* Corresponding Author's Email: recastro@mail.ru.

concentration of ~ 10^{16} cm^{-3} was found in $Ge_{28.5}Pb_{15}S_{56.5}$ system by thermoactivation spectroscopy methods in the temperature range $T = 260 ... 280K$ [5]. The objective of this study was to identify specific features of low-frequency relaxation processes in thin layers of $Ge_{28.5}Pb_{15}S_{56.5}$ glassy system by the dielectric spectroscopy (DS) method.

The DS method provides us a tool to discover specific features of polarization processes and their relationships with structural features of the studied material [6-8]. In this method the basic operation is to create a sine wave at the frequency of interest, apply it to the sample and measure the sample voltage $U(t)$ and result current $I(t)$. From this, the amplitude I_0 and phase angle φ of the current harmonic base component $I^*(\omega)$ is calculated by complex Fourier transform (FT) of $I(t)$. In addition to the phase detection, the FT suppresses all frequency components in $I(t)$ except a narrow band centered around the generator frequency. This improves accuracy and reduces noise and DC drifts by several orders of magnitude. E.g., measuring a signal covered by a noise signal of 1000 times larger is possible. Finally, the impedance $Z(\omega)$ and material parameters $\varepsilon^*(\omega)$ and $\sigma^*(\omega)$ are calculated [9].

EXPERIMENTAL SECTION

The elemental composition of samples was studied with Carl Zeiss EVO 40 scanning electron microscope (SEM). The maximum resolution of the microscope is 3 nm. To identify the atomic composition of $Ge_{28.5}Pb_{15}S_{56.5}$ films, the atomic content spectra were identified in the selected regions of the test samples (Figure 1). Data on the weight and atomic percentages of chemical elements in the samples are shown in Table 1. Dielectric spectra were evaluated on the films of $Ge_{28.5}Pb_{15}S_{56.5}$ glasses prepared by vacuum thermal evaporation. The samples had a sandwich configuration with aluminum electrodes and a contact area of 14.0 mm^2. The thickness of $Ge_{28.5}Pb_{15}S_{56.5}$ layers was measured with the ELF Spectroellipsometer and made ~ 2.0 μm. Dielectric spectra of the studied layers were measured with the "Concept-81" spectrometer (Novocontrol Technologies GmbH) designed for studying the dielectric and electrical properties of an extensive class of materials. Measurements were taken in the frequency range $f = 10^{-2}...10^{5}$ Hz and the temperature range $T = 273... 313K$. The voltage $U = 10^{-1}$ V was applied to the samples.

Spectra of the complex dielectric permittivity were calculated from the impedance spectra, according to the following formula:

$$\varepsilon^* = \varepsilon' - i\varepsilon'' = \frac{-i}{\omega Z^*(\omega)} \frac{1}{C_0} \qquad (1)$$

where, $C_0 = \varepsilon_0 S/d$ is the capacity of the empty cell.

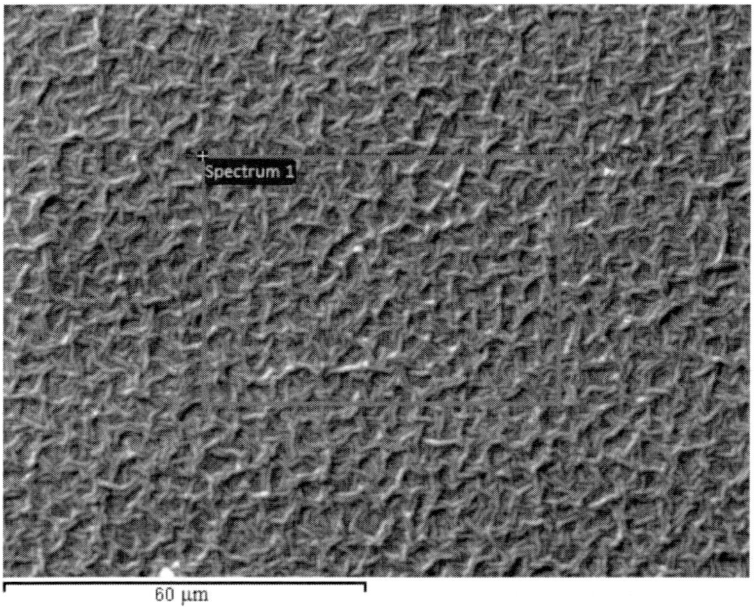

Figure 1. Image of the sample surface at a resolution of 60 μm.

Table 1. Elemental composition of $Ge_{28.5}Pb_{15}S_{56.5}$ samples (weight % and atomic % content of chemical elements)

Element	Nominal concentration	Mass fraction	Weight %	Weight % Sigma	Atomic fraction %
S K	24.80	0.9435	27.29	0.16	56.48
Ge K	32.68	0.9958	34.07	0.24	31.15
Pb M	31.38	0.8428	38.65	0.27	12.38
Total			100.00		

To determine the values of the system relaxation parameters, derived dielectric spectra were approximated by the Havriliak-Negami (HN) two-parametric empirical function [10] using the Novocontrol Winfit software. Based on these approximations, positions of the dielectric loss maxima were identified and the HN parameters were determined for the studied relaxation processes:

$$\varepsilon^*(\omega) = \varepsilon_\infty + \frac{\Delta\varepsilon}{\left[1 + (i\omega\tau)^{\alpha_{HN}}\right]^{\beta_{HN}}} \tag{2}$$

where, ε_∞ is the high-frequency limit of the real part of the dielectric permittivity, $\Delta\varepsilon$ is the dielectric increment (the difference between the low-frequency and high-frequency ε limits), $\omega = 2\pi f$, α_{HN} and β_{HN} are shape parameters that describe the symmetric ($\beta = 1$, the

Cole-Cole distribution) and asymmetric (α = 1, the Cole-Davidson distribution) expansion of the relaxation function, respectively.

RESULTS AND DISSCUSSION

Measurements of the dielectric loss tangent tanδ in $Ge_{28.5}Pb_{15}S_{56.5}$ system layers in the frequency range 10^{-1} ... 10^5 Hz at different temperatures (Figure 2) revealed the existence of the loss maximum, which shifts to higher frequency areas with increasing temperature. The presence of maxima on the tanδ curve at a relatively low frequency and temperature indicates the existence of the relaxation process responsible for relaxation losses in the samples. The presence of the relaxation process is also evidenced by the maximum on the temperature-dependent dielectric loss curves in the low-frequency regions.

In many dielectrics, relaxation processes are associated with the existence of a set of relaxation times, rather than a single relaxation time. These cases can be viewed as the presence of distributed relaxation times, and, consequently, of activation energies. Such distribution can be associated with the manifestation of relaxation processes of various nature or with the distribution over dipole concentration within the structure. In the case of ionic hopping processes, it is assumed that the potential energy changes after each hop and certain time is needed to return to the minimum potential energy. When the contribution of multiple mobile defects is considered, we obtain a set of relaxation times.

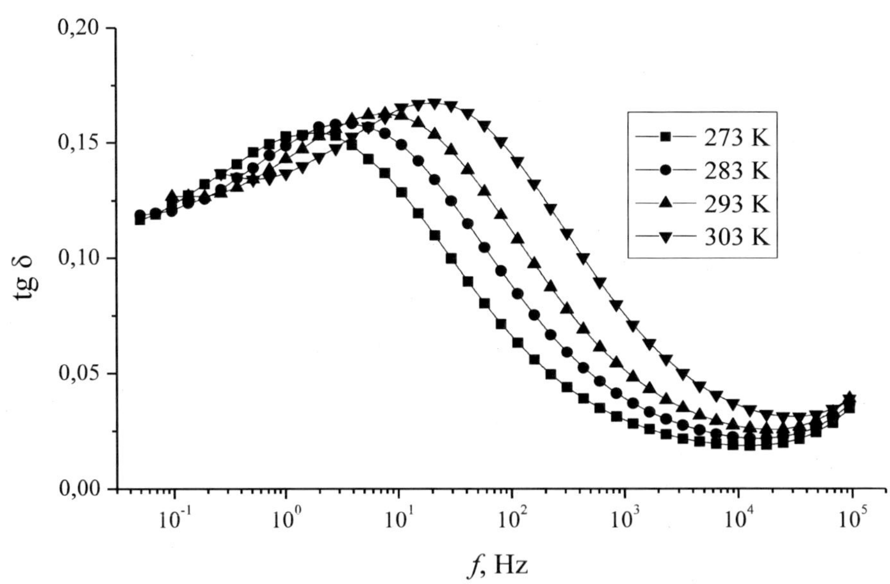

Figure 2. Frequency dependence of the dielectric loss (tanδ) at different temperatures

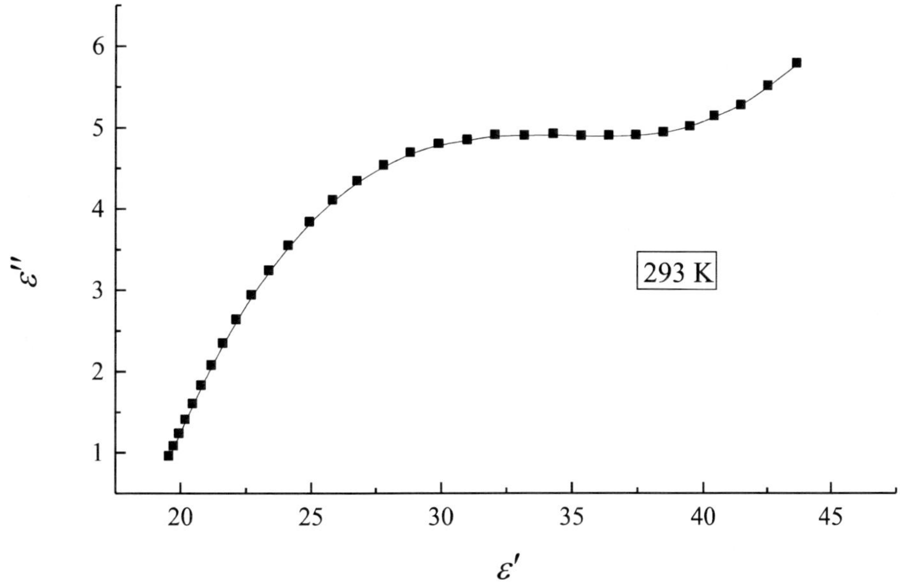

Figure 3. The Cole-Cole diagram for $Ge_{28.5}Pb_{15}S_{56.5}$ samples at room temperature.

In order to prove the existence of the "non-Debye" dispersion mechanism in the studied system, the Cole-Cole diagram ($\varepsilon'' = f(\varepsilon')$ was constructed. In our case, the existence of distribution of relaxators over time is manifested in the deviation from the hemispherical dependence with the radius of the hemisphere ($\varepsilon_s - \varepsilon_\infty / 2$) (Figure 3).

The shape of the dielectric loss curve of the loss factor ε''=(f) or the dielectric loss tangent tanδ = (f) can be used for estimating the distribution of relaxation times. The values of relaxation parameters derived from HN approximations of the experimental curves (2) also confirm the existence of distribution of relaxators, in line with the Cole-Cole model for the case of symmetric distribution of relaxation times. The temperature dependence of the frequency (relaxation time) with the maximum observed losses f_m (τ_M) allows us to determine the experimental activation energy E_a, i.e., the energy barrier to dipole orientation. The activation energy calculated from the temperature dependence for the most probable relaxation time (Figure 4) was equal to $E_p = (0.40 \pm 0.01)$ eV.

The observed patterns in the temperature and frequency dependences of dielectric losses in the low-frequency region can be explained by the original model suggested by the authors [11]. According to this model, the structure of chalcogenide glasses represents a set of dipoles formed by the charged defects of D^+ and D^- types. Each dipole has its specific relaxation time, which depends on the activation energy required for charge carriers to overcome the potential barrier. The existence of a potential barrier is determined by the Coulomb interaction between the neighboring defect states that form a dipole.

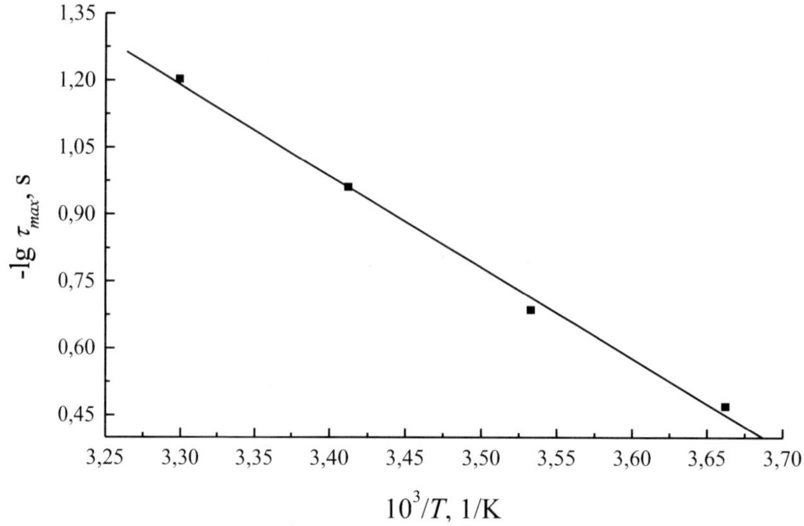

Figure 4. Temperature dependence for the most probable relaxation time in $Ge_{28.5}Pb_{15}S_{56.5}$ layers.

According to the CBH model (correlated barrier hopping model) [12], the value of potential barrier is defined:

$$W = W_M - \frac{ne^2}{\pi\varepsilon\varepsilon_0 r} \tag{3}$$

where W_M is the maximum barrier height, ε – the dielectric constant of the material, r – distance between two hopping sites, and n is the number of the electrons involved in a hop (n = 1 and 2 for the single polaron and bipolaron processes respectively).

The findings of our study correlate with the conclusions of the authors [5] and support the hypothesis of the existence of a defect states spectrum, when the electron exchange between the states creates quasi-dipoles responsible for the dielectric permittivity dispersion and dielectric losses in the low-frequency region. The electron exchange can occur, for example, between germanium atoms that are present in both bivalent and tetravalent states in the glass structure [13].

CONCLUSION

As can be seen from the above, our study has revealed the existence of the dielectric loss maximum, which indicates the existence of the relaxation process. The values received for relaxation parameters confirm the existence of distribution of relaxators, in line with the Cole-Cole model for the case of symmetric distribution of relaxation times. The activation energy of dielectric relaxation process was equal to E_p = (0.40 ± 0.01) eV. The observed patterns are explained by the model, according to which the structure of chalcogenide glasses represents a set of dipoles formed by the charged defects of D^+ and D^- types.

ACKNOWLEDGMENTS

The reported study was supported by the Ministry of Education and Science of the Russian Federation (project № 3.5005.2017/BY).

REFERENCES

[1] Kolobov, A. V.; Fons, P.; Tominaga, J.; Uruga, T.; Haines, J. *Proceedings of the European Phase Change and Ovonics Symposium;* Cambridge: United Kingdom, 2005.

[2] Kukreti, A. K.; Gupta, S.; Saxena, M.; Rastogi, N. *International Journal of Innovative Research in Science, Engineering and Technology*. 2015, vol. 4, pp. 18609-18614.

[3] Bekheet, A. E.; Hegab, N. A. *Vacuum*. 2008, vol. 83(2), pp. 391-396.

[4] Anderson, P. W. *Physical Review Letters*. 1975, vol. 34(15), pp. 953.

[5] Bordovsky, G. A.; Bordovsky, V. A.; Anisimova, N. I.; Castro, R. A.; Seldjaev, V. *Abstr. of the II Intern. Materials Symp.* 2003, p. 59.

[6] Avanesyan, V. T., Bordovskii, G. A., & Castro, R. A. *Glass Physics and Chemistry*. 2000, vol. 26(3), pp. 257-259.

[7] Castro, R. A.; Bordovsky, G. A.; Bordovsky, V. A.; Anisimova, N. I. *Journal of Non-crystalline solids*. 2006, vol. 352(9-20), pp. 1560-1562.

[8] Castro, R. A.; Bordovsky, V. A.; Grabko, G. I. *Glass Physics and Chemistry*. 2009, vol. 35(1), pp. 43-46.

[9] Schaumburg, G. *Dielectric Newsletter*. 1999 (5), pp. 4-6.

[10] Kremer, K. and A. Schonhals (Eds.). *Broadband Dielectric Spectroscopy*, Springer, Berlin Heidelberg. 2003. –729 c.

[11] Elliott, S. R. *Advances in Physics*. 1987, vol. 36(2), pp. 135-217.

[12] Giuntini, J. C.; Zanchetta, J. V.; Jullien, D.; Eholie, R.; Houenou, P. *Journal of Non-Crystalline Solids*. 1981, vol. 45(1), pp. 57-62.

[13] Bordovskii, G. A.; Castro, R. A. *Glass Physics and Chemistry*. 2006, vol. 32(3), pp. 315-319.

In: Applied Aspects of Nano-Physics …
Editors: Kirill Levine and Andrey Syrkov
ISBN: 978-1-53614-708-7

Chapter 36

DIELECTRIC RELAXATION IN THIN LAYERS OF THE AROMATIC THERMOPLASTIC POLYIMIDE

Natalia A. Nikonorova*[1]*, Aleksei A. Kononov*[2]*,
***Hong T. Dao*[2] *and Rene A. Castro*[2,*]**
[1]Institute of Macromolecular Compounds
of Russian Academy of Sciences, Saint Petersburg, Russia
[2]Herzen State Pedagogical University of Russia,
Saint Petersburg, Russia

ABSTRACT

The dielectric behavior of the ODPA-OOD aromatic thermoplastic polyimide was studied. Three regions of dipole polarization relaxation were identified in dielectric spectra of all samples: γ, β (glassy state) and α (rubbery state). γ process is determined by the local mobility of phenylene rings in the diamine part of a macromolecule. β process represents a superposition of several molecular mobility modes with similar relaxation times that reflect local mobility of phenylene rings in the diamine and dianhydride parts of a macromolecule and the adjacent polar groups. The molecular source of α process is presented by large-scale segmental mobility of the macromolecule ridge; the major contribution to intermolecular interactions is provided by dispersion forces between planar phenylene rings.

Keywords: dielectric relaxation, thermoplastic aromatic polyimide, glass transition temperature

* Corresponding Author's Email: recastro@mail.ru.

INTRODUCTION

Polyimides (PIs) feature good physical and chemical properties, a stable structure, high thermal and chemical resistance, and are therefore widely used as insulating materials in aircraft and space engineering, where extreme conditions are common, specifically high temperatures (up to 400°C) and irradiation. The electrical characteristics of polyimides, such as high specific resistivity ($\rho_v \sim 10^{15}$ Ohm*m), high dielectric strength (Ep ~300 MV/m), and relatively low dielectric permittivity ($\varepsilon' \sim 2.5...3.5$) are advantageous for PI applications in electronic devices, for example, as electret materials [1, 2]. PIs are actively used as separation membranes with a high rate of gas permeability and selectivity [3-5]. Their synthesis methods are well-developed and enable manufacturing PIs of the most diverse chemical structure and hence, with highly variable physical properties [1]. Thermoplastic aromatic PIs are of special interest, since they are easily processed and can provide the basis for nanocomposites of new generation with improved mechanical, electrical and membrane properties.

PIs have been studied by different methodology including relaxation methods (dielectric, mechanical, and nuclear magnetic resonance). Relaxation methods provide diversified information about the structure and molecular mobility of a polymer at all levels of the molecular organization. Thus, regions of the maximum dielectric loss (processes of dipole polarization relaxation) can be observed on the dielectric spectra; these are determined by both the local molecular mobility of certain groups, and the cooperative movement of large kinetic segments and even of the macromolecule as a unit. Relaxation time and activation characteristics of molecular motion depend on molecular interactions, which are determined by the chemical structure and intermolecular interactions [6, 7].

At least three dipole polarization relaxation regions were found in PIs of different structure: γ, β and α processes as the temperature increased γ, β relaxation processes occur in the glassy state, and α process was supposed to be associated with crossing the glass transition temperature, i.e., with the segmental large-scale mobility of the main chains [8-14].

The objective of this study was to identify the features of dielectric γ, β and α relaxation processes, the relationship between molecular mobility and the chemical structure of the thermoplastic polyimide under test.

EXPERIMENTAL SECTION

Under this study, molecular mobility of the thermoplastic polyimide ODPA-OOD (Figure 1) was examined by dielectric spectroscopy method.

Polyimide was manufactured via two-stage synthesis. The first stage (polycondensation reaction) – obtaining polyamide acid (PAA) – was performed in equimolar solution of aromatic dianhydride and aromatic diamine in dimethylacetamide, at room temperature. Films were poured from PAA solution onto the glass. The second stage (thermal treatment of PAA films) resulted in PI production. The synthesis details and the method of producing PI films are described in [15, 16]. Dielectric measurements were taken from films pre-heated at 300^{0}C for 30 minutes.

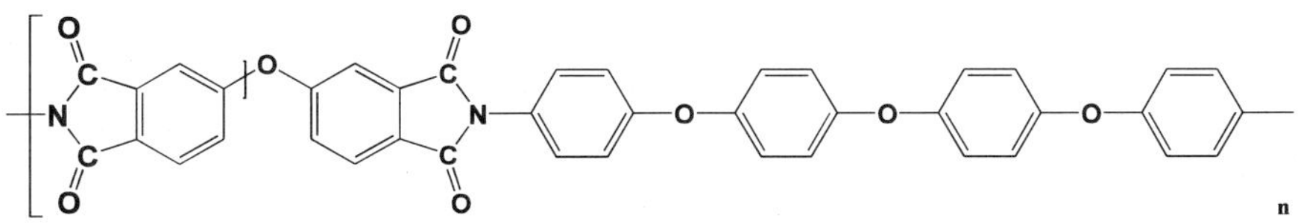

Figure 1. Structure of ODPA-OOD thermoplastic polyimide.

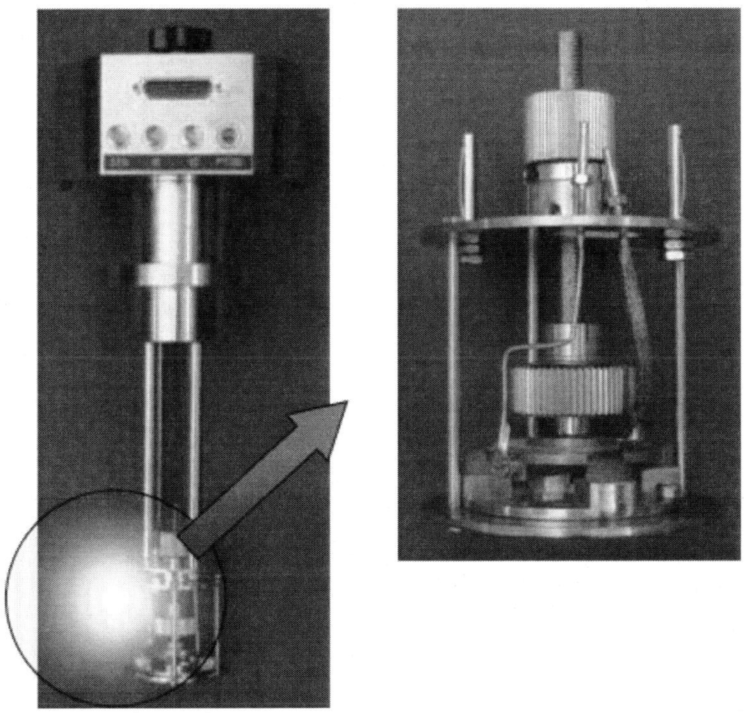

Figure 2. ZGS Alpha active sample cell.

Dielectric spectra were obtained with "Concept 81" broadband dielectric spectrometer (Novocontrol Technologies GmbH), with ALPHA-ANB automatic high-resolution frequency spectrum analyzer. Films of 25...40 μm thickness, compressed between brass electrodes (diameter of the upper electrode – 20 mm) at a temperature of ~30^{0}C above the glass transition temperature, were taken as samples (Figure 2). Temperature-frequency dependences of dielectric permittivity ε,' dielectric loss factor ε’,’ dielectric loss tangent tgδ, and conductivity σ were obtained for all PI samples in the frequency range 10^{-1}Hz ...10^{6}Hz and the temperature range 173K...623K.

The spectra of complex dielectric permittivity and complex conductivity were calculated from the impedance spectra:

$$Z^*(\omega) = R + \frac{1}{i\omega C} = Z' + iZ'' = \frac{U_0}{I^*(\omega)} \tag{1}$$

according to the formulas:

$$\varepsilon^* = \varepsilon' - i\varepsilon'' = \frac{-i}{\omega Z^*(\omega)} \frac{1}{C_0} \tag{2}$$

$$\sigma^* = \sigma' - i\sigma'' = \frac{-i}{\omega Z^*(\omega)} \frac{S}{d} \tag{3}$$

where $C_0 = \varepsilon_0 S/d$ is the capacity of the empty cell.

Dielectric spectra were analyzed with the Havriliak-Negami (HN) two-parametric empirical function [10] using the Novocontrol Winfit software. Based on these approximations, positions of the dielectric loss maxima were identified and the HN parameters were determined for the studied relaxation processes:

$$\varepsilon^*(\omega) = \varepsilon_\infty + \frac{\Delta\varepsilon}{\left[1 + (i\omega\tau)^{\alpha_{HN}}\right]^{\beta_{HN}}}, \tag{4}$$

where, ε_∞ is the high-frequency limit of the real part of dielectric permittivity, Δε is the dielectric increment (the difference between the low- and high-frequency limits), ω = 2πf, α_{HN} and β_{HN} are shape parameters that describe the symmetric (β = 1, the Cole-Cole distribution) and asymmetric (α = 1, the Cole-Davidson distribution) expansion of the relaxation function, respectively. The most probable relaxation time corresponding to the dielectric loss maximum was evaluated by the formula:

$$\tau_{макс} = \tau_{HN}\left[\frac{\sin\left(\frac{\pi(\alpha_{HN})\beta_{HN}}{2(\beta_{HN}+1)}\right)}{\sin\left(\frac{\pi(\alpha_{HN})}{2(\beta_{HN}+1)}\right)}\right]^{1/(\alpha_{HN})} \tag{5}$$

Results and Discussion

Temperature and frequency dependences of tgδ and ε" were essentially similar for all tested samples. To visualize the dielectric behavior over the entire temperature range, the temperature dependence of tgδ is presented in Figure 3. This dependence shows three regions of the maximum within the studied temperature and frequency range, which are determined by the processes of dipole polarization relaxation, since the position of $tg\delta_{макс}$ is shifted towards higher temperatures against the change in frequency. These regions of the maximum are indicated (in the order of temperature increase) as γ, β (the glassy state), and α (the rubbery state). At temperatures above α process, another $tg\delta_{max}$ region was observed, which corresponded to the DC conductivity relaxation (not displayed in the figure).

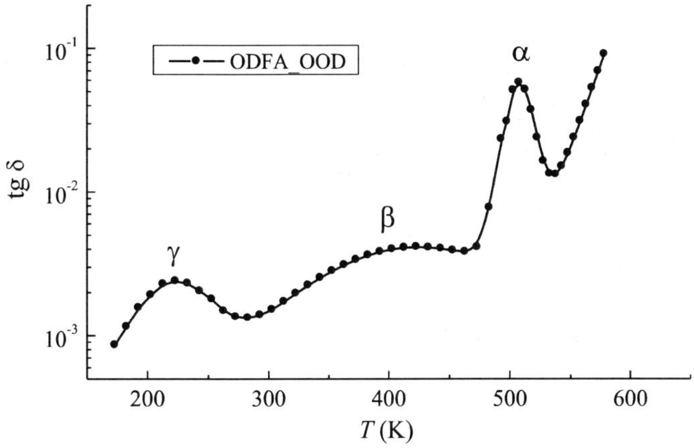

Figure 3. Temperature dependence of tgδ at frequency f = 10^3 Hz.

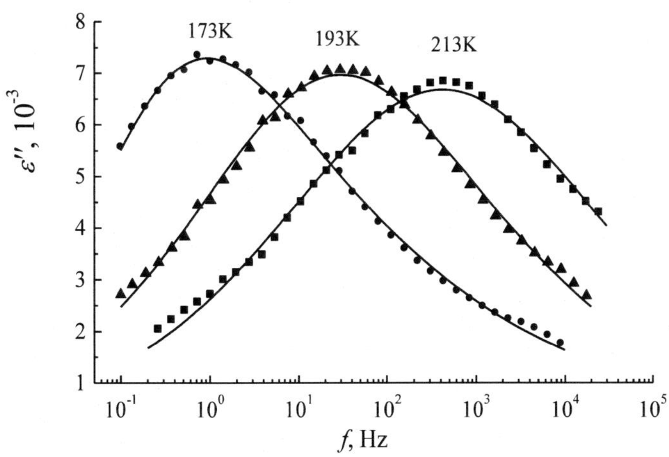

Figure 4. Frequency dependence of loss factor in γ-process region at different temperatures. Solid lines refer to the HN approximation of the experimental curves.

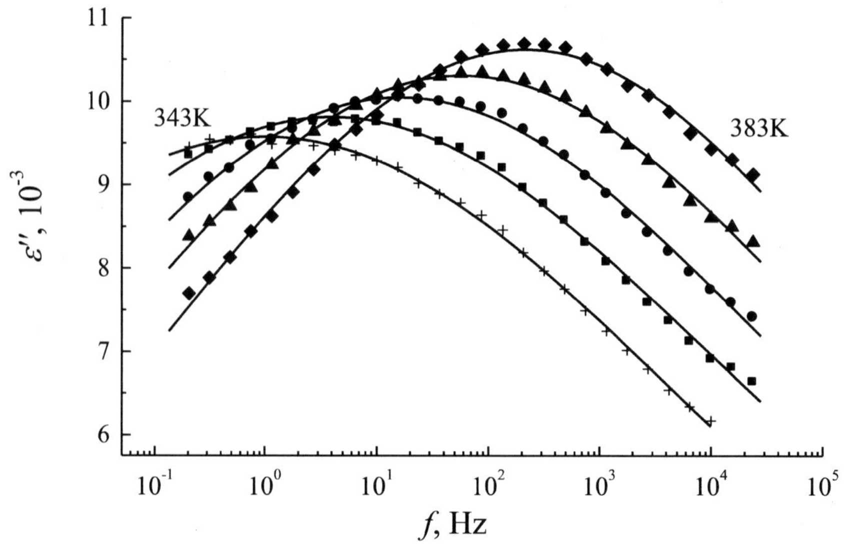

Figure 5. Frequency dependence of dielectric loss factor in β-process region at different temperatures. Solid lines refer to the HN approximation of the experimental curves.

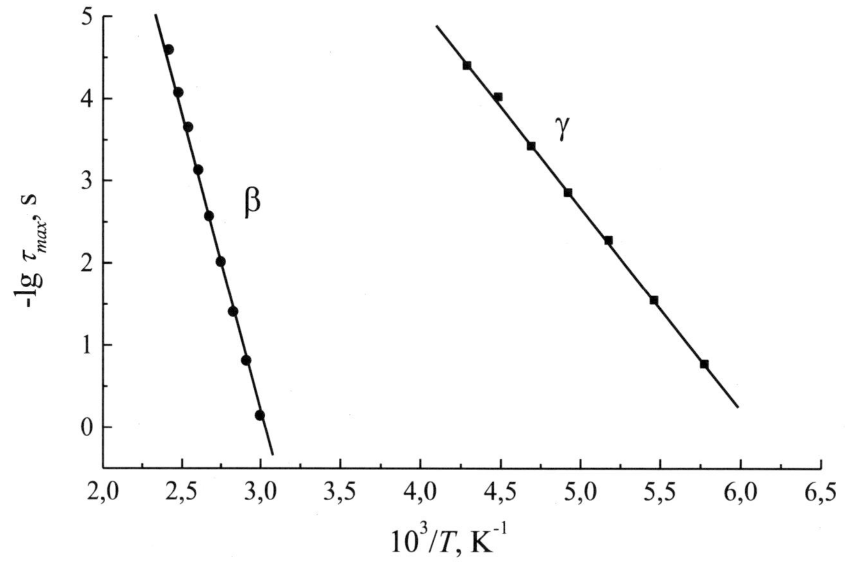

Figure 6. Temperature dependence of the relaxation time calculated by the HN formula in the region of γ and β processes.

Figure 4, 5 demonstrate the dependences of dielectric loss factor in the temperature regions corresponding to γ, β relaxation processes. The values of the relaxation times and their temperature dependence calculated by the HN formula (2) are presented in Figure 6 and Table 1.

Table 1. Parameters of the Arrhenius equation (6) for β and γ processes

Sample	-logτ$_0$, c	E_a, kcal/mol	-logτ$_0$, c	E_a, kcal/mol
	β- process		γ- process	
ODPA-OOD	23.02	34.90	14.98	11.25
Error (±)	0.11	0.25	0.13	0.12

Dependences $-\lg \tau_{max} = f(1/T)$ for both processes can be described by the Arrhenius equation:

$$\tau(T)_{max} = \tau_0 \exp\left(\frac{E_a}{RT}\right) \tag{6}$$

where, $\tau_0 = \tau_{max}$ at $T \rightarrow \infty$, E_a is the activation energy, R is the universal gas constant (R = 8.314 J/mol·K). The linearity of $-\lg \tau_{max} = f(1/T)$ dependence is typical for the local forms of molecular mobility described by the Debye model. This model assumes the absence of intermolecular interactions and here the activation energy does not depend on temperature. The equation parameters (6) for these processes are presented in Table 1.

According to the Debye model, a charged particle (dipole) bounded by the potential barrier E_a = 8…12 kcal/mol oscillates with a frequency $f_0 = 10^{12}…10^{13}$ Hz ($-\lg \tau_0$ = 13…14 s). Data from Table 1 allow us to conclude that parameters of the equation (6) for γ-process conform to the characteristics of the local process. At the same time, -logτ$_0$ and E_a values for all samples are greater than those predicted for the local mobility by the Debye theory. This fact can be explained if we assume the presence of intermolecular interaction.

The same as for the PIs tested in this study, the existence of three relaxation modes, α, β and γ, was discovered earlier in linear polyesters and polyimides of various structure. The authors of these works [8-14] suggested molecular mechanisms of these processes.

Several explanations were proposed for high-frequency γ process. Specifically, it was demonstrated that this process was determined by the local mobility of ether groups adjacent to the flexible parts and bonded water molecules. At the same time, the intensity of γ process correlated with the bound water content [10, 12, 14]. As a rule, $tg\delta_{max}$ temperature (at 1 Hz) in γ process was independent of the PI structure and made ~173K. In our case, we can suggest that the molecular mechanism of the fastest γ process is determined by the mobility of phenylene rings with adjacent ether groups in the diamine part of a PI macromolecule. Dielectric losses of a PI in glassy state are small for γ process and in the region of tgδ maximum stay within $2.10^{-3}…7.10^{-3}$ (Figure 2).

The emergence of β processes was mainly attributed to mobility in the dianhydride part of a macromolecule [10], rotation of para-phenylene units in the diamine part [8], or correlated local movements in both parts [8, 10, 12, 14, 18-20]. In certain cases, β process can be subdivided into β_1 and β_2 processes [8, 11], where each reflects mobility of the dianhydride or diamine PI component. In β process, $tg\delta_{макс}$ temperature (at 1 Hz) in γ process depended on the chemical structure, thermal history and PI morphology, and stayed within 323K...423K. In the case of the PI system (our study), kinetic units responsible for the emergence of higher-temperature (low-frequency) β process should be more extended than those for γ process. Considering the structure of the tested PIs, we can suggest that β process is determined by local movements of the phenylene rings in the diamine and dianhydride parts of a macromolecule with adjacent ether, amide or sulphone groups. That is, β process represents a superposition of several molecular mobility modes with similar relaxation times, impossible to get separated, unlike those presented in recent study results [8, 12]. For this PI series, the temperature-frequency coordinates of β process (relaxation times) and its intensity are virtually independent of the chemical structure, and $tg\delta_{max}$ temperature makes ~ 348K.

Figure 7 presents the frequency dependence of dielectric loss in the temperature region corresponding to α relaxation processes. Contributions of dielectric loss and conductivity to the general picture of dielectric relaxation are indicated in Figure 8.

The values of relaxation time and their temperature dependence calculated by the HN formula (2) are shown in Figure 9. Here, $lg\tau_{max} = \varphi(1/T)$ temperature dependence of the tested samples reveal a nonlinear pattern in α process region (Figure 9). This pattern is typical for cooperative relaxation processes in molecular motion, which are characterized by a wide set of relaxation times and are implemented as a joint correlated movement of a large number of segments belonging to neighboring macromolecules (Table 2).

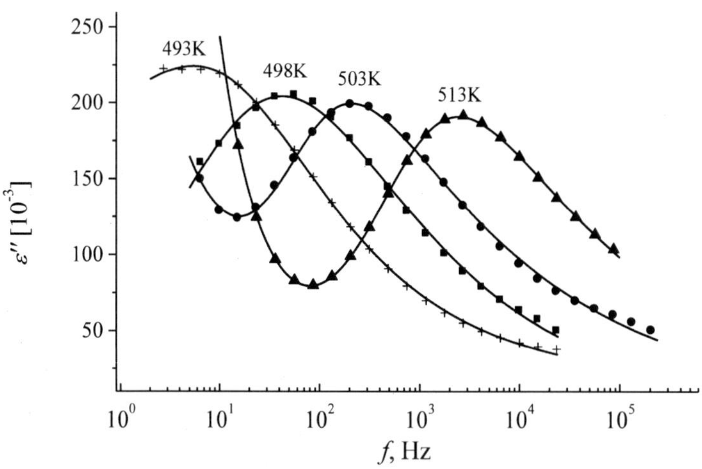

Figure 7. Frequency dependence of dielectric loss in α process region at different temperatures. Solid lines refer to the HN approximation of the experimental curv.

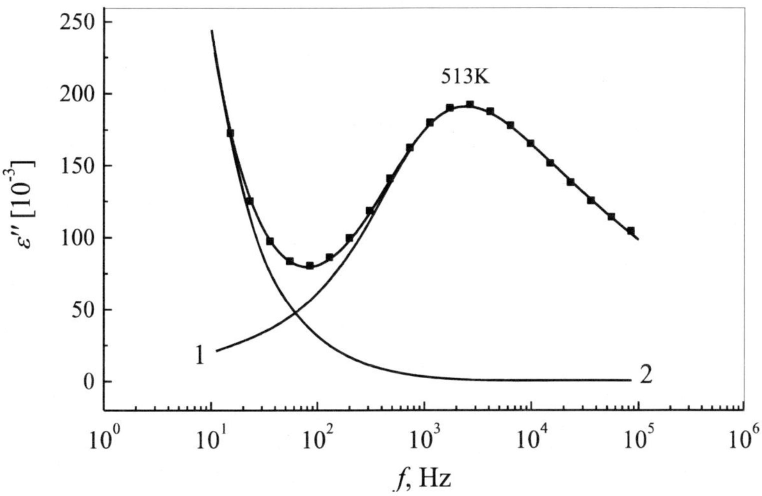

Figure 8. Frequency dependence of dielectric loss in α process region. Solid line refers to the HN approximation of the experimental point (1 –dielectric loss; 2 – conductivity).

The effect is that the molecular mobility of kinetic segments depends on the state of the immediate environment and is determined largely by intermolecular interactions.

The activation energy of this cooperative process depends on the temperature and $\lg\tau_{max}=\varphi(1/T)$ dependences are well described by the empirical Vogel–Tammann–Fulcher (VTF) equation [17]:

$$\tau_{MAKC} = \tau_0 \exp(\frac{B}{T-T_0}) \quad (7)$$

where, τ_0, B and T_0 are temperature-independent parameters. T_0 is the so-called Vogel temperature. B parameter represents a measure of cooperativity of the relaxation process. The smaller B is, the greater are distortion and deviation from linearity of $-\lg\tau_{max} = \varphi(1/T)$ dependence, as well as the cooperativity of the process (parameters of the equation (7) for α process in the system under study are presented in Table 2).

For this thermoplastic PI, the same as for the PIs of various structure studied previously, α process can be explicitly associated with crossing the glass transition temperature. The molecular source of α process is presented by large-scale segmental mobility of the macromolecule ridge. The temperature-frequency coordinates of α process (Figure 9), dependences $-\log\tau_{max} = \varphi(1/T)$, separate the glassy state region of the polymer (right) from the rubbery state region (left). The glass transition temperature, T_c (see Table 2), was determined by extrapolation of $\log\tau_{max} = \varphi(1/T)$ dependence described by the VTF equation to $-\log\tau_{max}= 0$ ($\tau_{max} = 1$ s). For linear polymers, T_c is determined primarily by intermolecular and dipole-dipole interactions between macromolecules. In the case of PIs, the major contribution to intermolecular interactions is probably provided by dispersion forces between planar phenylene rings.

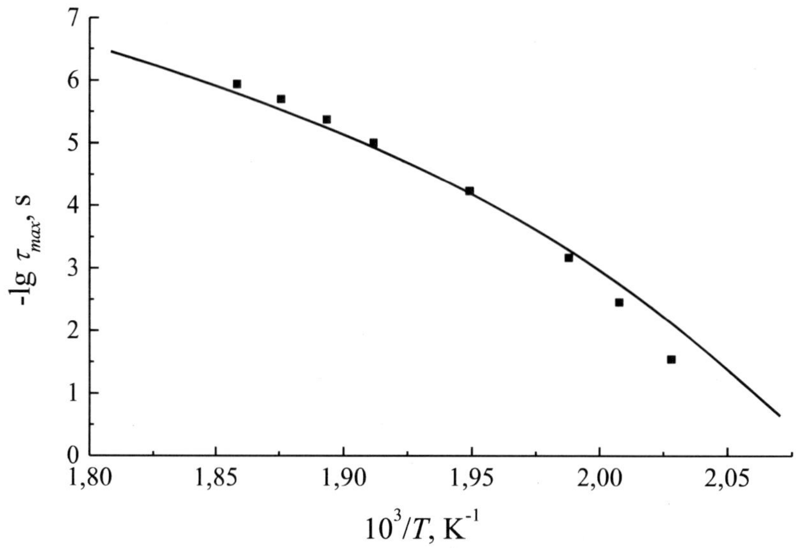

Figure 9. Temperature dependence of the relaxation time calculated by the HN formula in α process region.

Table 2. Parameters of the equation (6) in α process region for ODPA-OOD

Sample	$-\lg\tau_0$ [c]	B, K	T_0, K	T^*_c, ^{0}C
ODPA-OOD	11.6	1564	421	480
Error (±)	0.35	46.92	16.84	9.6

*at $\lg\tau_{max} = 0$.

Conclusion

Molecular mobility of the ODPA-OOD aromatic thermoplastic polyimide with a constant chemical structure of diamine and dianhydride parts of a macromolecule was studied with dielectric method. Dielectric spectra of all PI samples in the examined temperature and frequency range were essentially similar and revealed three regions of dipole polarization relaxation (in rising temperature order): γ, β (glassy state) and α (rubbery state). The empirical HN formula was applied to quantify γ and β relaxation processes. A comparison between the dielectric behaviors of similarly structured polymers and this polyimide allows for identifying relaxation processes.

The fastest γ process can be linked to the local mobility of phenylene rings in the diamine part of a macromolecule, whereas β process represents a superposition of several molecular mobility modes with similar relaxation times and can be associated with the mobility of phenylene rings in the diamine and dianhydride parts of a macromolecule and the adjacent polar groups.

The temperature dependence $\log\tau_{max} = \varphi(1/T)$ in α process region displays a nonlinear pattern, which is typical for cooperative relaxation processes in molecular motion. In the case of the studied system, the major contribution to intermolecular interactions is probably provided by dispersion forces between planar phenylene rings.

ACKNOWLEDGMENTS

The reported study was supported by the Ministry of Education and Science of the Russian Federation (project № 3.5005.2017/BY).

REFERENCES

[1] Bryant, R. G. *Polyimides,* John Wiley & Sons: New York, 2002.

[2] Sroog, C. E. In *Encyclopedia of Polymer Science and Technology, 1st edition*, John Wiley & Sons, Inc. 1969, vol. 11, pp. 247-272.

[3] Kase, Y. *Advanced Membrane Technology and Applications*. 2008, pp. 581-598.

[4] Matsumoto, K., Xu, P. *Journal of Membrane Science*. 1993, vol. 81, № 1-2, pp. 23-30.

[5] Matsumoto, K., Xu, P., Nishikimi, T. *Journal of Membrane Science*. 1993, vol. 81, № 1-2, pp. 15-22.

[6] McCrum, N. G., Read, B. E., Williams, G. *Anelastic and dielectric effects in polymeric solids,* John Wiley and Sons: London, 1967.

[7] Hedvig, P. *Dielectric Spectroscopy of Polymers*. 1977, pp. 18-22.

[8] Sun, Z., Dong, L., Zhuang, Y., Cao, L., Ding, M., Feng, Z. *Polymer.* 1992, vol. 33, pp. 4728-4731

[9] Coburn, J. C., Soper, P. D., Auman, B. C. *Macromolecules.* 1995, vol. 28, pp. 3253-3260.

[10] Havriliak, S., Negami, S. *Polymer.* 1967, vol. 8, pp. 161-210.

[11] Cheng, S. Z. D., Chalmes, T. M., Gu, Y., Yoon, Y., Harris, F. W., Cheng, J., Fone, M., Koenig, J. L. *Macromolecular Chemistry and Physics.* 1995, vol. 196, pp. 1439-1451.

[12] Chisca, S, Musteata, V. E., Sava, I., Bruma, M. *European Polymer Journal.* 2011, vol. 47, pp. 1176-1197.

[13] Jacobs, J. D., Arlen, M. J., Wang, D. H. *Polymer.* 2010, vol. 51, pp. 3139-3146.

[14] Bas, C., Pascal, T., Alberola, N. D., *Polymer Engineering & Science.* 2003, vol. 43, pp. 344-355.

[15] Belkevich, N. G., Svetlichnyi, V. M., Kourenbin, O. I., Milevskaya, I. S., Nesterov, V. V., Kudryavtsev, V. V., Frenkel, S. Ya. *High-Molecular Compounds.* 1995, vol. A37, № 8, pp. 1357-1360.

[16] Sroog, C. E. *Progress in Polymer Science.* 1991, vol. 16, pp. 561.

[17] Kremer, K., Schonhals, A., *Broadband Dielectric Spectroscopy*, Springer: Berlin, 2003.

[18] Qu, W., Ko, T. M., Vora, R. H., Chung, T. S. *Polymer.* 2001, vol. 42, pp. 6393-6401.

[19] Burshtein, L. L., Borisova, T. I., Zhukov, S. V., Nikonorova, N. A., Skorokhodov, S. S. *Polymer.* 1999, vol. 40, pp. 1881-1887.

[20] Nikonorova, N. A., Balakina, M. Yu., Fominykh, O. D., Pudovkin, M. S., Vakhonina, T. A., Diaz-Calleja, R., Yakimansky, A. V. *Chem. Phys. Lett.* 2012, vol. 552, pp. 114-121.

In: Applied Aspects of Nano-Physics …
Editors: Kirill Levine and Andrey Syrkov
ISBN: 978-1-53614-708-7

Chapter 37

REACTIVITY OF HYDROXYL GROUPS ON THE SURFACE OF SOLIDS

*Yu. K. Ezhovskii**
Saint Petersburg State Technological Institute (Technical University),
Saint Petersburg, Russia

ABSTRACT

The concept of the macromolecular structure of solids and quantitative estimation of surface reactivity using inductive constants is applied to the estimate of surface nanostructures on dispersed and single crystal matrices. The possibilities of the proposed approach are demonstrated by the example of the synthesis of oxide nanostructures on a silicon surface.

Keywords: reactivity of solid surfaces, dispersed and single crystal matrices, oxide nanostructures

INTRODUCTION

The development of the chemical nanotechnology of low dimensional systems on a solid surface are mainly associated with the possibility of predicting the conditions of reactions using quantitative estimation of their reactivity. One such approach is a correlation analysis system based on the principle of the linearity of free energies and is successfully used in the chemistry of high molecular weight compounds [1]. In

* Corresponding Author's Email: ezhovski1@mail.ru.

accordance with this principle, the reactivity can be predicted within a series of isostructural compounds using empirical or experimentally determined constants.

According to the core models of the structure of solid materials and the chemistry of supramolecular compounds developed on its basis [2, 3], any chemical transformations on a solid surface can be considered as similar to those of a polymer. In this case, an approach based on application of the principle of the linearity of free energies and a correlation-analysis system in the chemistry of a solid surface can become the basis for the creation of a quantitative theory of surface reactions.

In this work we considered some results indicating the prospects of applying such an approach in studying the chemistry of a solid surface and, in first and foremost, for quantitative estimation of the reactivity of the functional groups of a surface during atomic layer deposition. The possibilities of the developed approach are demonstrated by the example of silicon oxide and single-crystal silicon.

THEORETICAL ANALYSIS

In accordance with the core concept of the structure of solids [2], any solid compound is a combination of a chemically inert core (X–) and reactive functional groups of the surface (Y) (e.g., –OH). All surface reactions with the participation of these groups not affecting the structure of the core can be considered as similar to polymer transformations [3]. The ability of the core as a substituent to any type of interaction is quantitatively characterized by the corresponding constant (σ_X), the numerical value of which has a relative character, i.e., makes it possible to estimate the effect of the given substituent with respect to the reference one. This makes it possible to apply the principle of the linearity of free energies for different solid compounds having similar-type functional groups within the formal approach and use linear correlations of the type:

$$A = A_{XY} + \alpha_Y \sigma_X, \quad (1)$$

where A is any correlated quantity (e.g., the velocity constant (K) or logK);

α is the proportionality coefficient reflecting the sensitivity of the reaction center to the effect of the substituent in the given reaction. Indices Y and X refer, respectively, to the reaction center (the functional group) and the solid core.

Qualitative interpretation of the reactivity of X_i–Y molecules lies in revealing the mechanism of the effect of Xi on the reaction center Y. It is established [1] that the main contribution introduces inductive, steric, resonance and other effects. The effective electroneg-ativity X_i usually characterized by the inductive constant σ_i, which reflects the total inductive effect and the polar resonance of this substituent. The most widespread concept of the inductive effect is intramolecular polarization caused by a charged or electronegative group and transferred along the bond chain.

The influence of the core of the solid, and, more exactly, the surface group of its atoms can be characterized by the value of the inductive constant (σ_S), which can be determined, e.g., from shifts of the frequencies of valence vibrations of the element–hydride bond, in particular, the Si–H bond in the functional group of the surface compound. This bond is very sensitive to the effect of substituents and the shift of its frequencies almost excludes the effect of steric factors. In this case Eq. (1) takes the form:

$$\nu_\iota = \nu_0 + \alpha\,(\Sigma\,\sigma_\iota + \sigma_\Sigma), \qquad (2)$$

where ν_i, ν_0 are the frequencies of the valence vibrations in the given compound and in the reference one, e.g., methyl-substituted silane or siloxane. If the Taft constants (σi) are used as the inductive constants σi characterizing the effect of other substituents (R_i) in the surface compound, the inductive constant will be also determined on the scale of the inductive Taft constants. The availability of information on the Taft constants for a large number of low-molecular-weight substituents [4] greatly simplifies the calculation part of the technique and makes the proposed approach quite universal.

RESULTS AND DISCUSSION

The coefficients of Eq. (2) for a reaction series of solid oxides were determined using samples of silicon dioxide (aerosil A-175) with different chemosorbed chlorinated silanes [5, 6]. As a result of mathematical processing of the experimental data (ν_i) obtained for a series of surface compounds and values of Taft constants (σ_i^*) for such substituents as Cl_2MeSiH ($\Sigma\sigma_i^*$=2.88), Cl_3SiH ($\Sigma\sigma_i^*$= 5.76), Cl_2PhSiH ($\Sigma\sigma_i^*$= 3.48) and $ClPh_2SiH$ (($\Sigma\sigma_i^*$= 1.10) [4], we determined the coefficients of the relation (2) and obtained the expression:

$$\nu_\iota = 2064 + 19.21\,(\Sigma\,\sigma_\iota^* + \sigma^*_\Sigma), \qquad (3)$$

The reactivity of the surface hydroxyl groups depends on the temperature of dehydroxylation of the silica surface, i.e., on their concentration. This is caused by an increase in the electronegativity of the surface silicon atoms due to the removal of hydroxyls and (or) water molecules from their coordination sphere [7] and redistribution of the electron density in the surface Si–O. Consequently, for the samples with different degrees of hydration, the numerical values σ_S^*should differ, quantitatively reflecting this phenomenon. The experimental results confirmed this assumption. The values σ_S^* were calculated using Eq. (3) for a series of aerosol A-175 amples preliminarily heat treated in vacuum in the temperature range of 473-1073K and having different concentrations of hydroxyl groups from the experimental data, according to the values ν_i (SiH) (see the table) [6] for chemosorbed dimethylchlorosilane. They can serve as a quantitative measure of the effect of a surface silicon–oxygen group (or even the whole surface) on the reaction center, e.g., the hydroxyl group, and characterize its activity. The results show that with an increase in temperature the values σ_S^* increase, and, consequently, the reactivity of the surface hydroxyls also increases.

Table 1. Effect of heat treatment on the concentration of hydroxyl groups and inductive constant σ^*_S for SiO_2 calculated according to νi(SiH) data for chemosorbed dimethylchlorosilane

T, K	473	573	673	773	873	973	1073
C_{OH}, μmole/m^2	7.8	5.6	3.4	3.0	2.6	2.3	1.5
ν_i, (SiH), cm^{-1}	2152	2155	2161	2163	2165	2168	2170
σ^*_S	4.57	4.73	5.05	5.15	5.25	5.41	5.52

A series of important parameters of surface processes, by the example of which it is possible to demonstrate the possibilities of the proposed approach, can be determined from the values of the inductive constants (σ_S^*). We consider, e.g., the surface processes on silicon and its dioxide. Silicon atoms are coupled with the Si–H center by the siloxane bond, which is capable of quenching inductive-resonance effects. The quenching coefficient (transmission factor of the bond Z) [8] can also be determined from the relation between the inductive constants for compounds with a bond chain:

$$\sigma_{(\equiv Si\text{-}O\text{-}Si\equiv)n}=\sigma_{(\equiv Si\text{-})}\cdot Z^n, \quad (4)$$

where n is the number of ≡Si–O–Si≡ bonds.

The numerical value of Z found from the values for a series of surface groups with a chain of siloxane bonds (n = 0, 1, 2, 3) $Z_{(Si–O–Si)}$ was 0,51±0,04, which does not differ to a great degree from the value for linear siloxanes (Z=0.48) [8]. This parameter can serve as a quantitative characteristic of the long-range effect of the silicon–oxygen core and be used, e.g., for estimating the reactivity of the functional group of a chain of siloxane bonds removed from the core or to take into account a change in its reactivity during synthesis of the different functional groups on the surface of matrices.

Application of the inductive effect in quantitative estimation of the reactivity of the surface of single-crystal, in particular semiconductor, matrices, makes it possible to predict the processes of the matrix synthesis of nanostructures on semiconductors. As an example, one can turn to the results of studies performed on silicon samples with the orientations (100), (110) and (111) by the method of multiple frustrated total internal reflection (FTIR) infrared (IR) spectroscopy [9, 10]. The study of the hydroxylation processes of freshly etched silicon and the subsequent sorption of dimethylchlorosilane (DMCS) showed that the silanol groups of the silicon surface do not participate in the reaction with DMCS, while the hydroxyl groups of the silica surface under similar conditions form thermostable surface compounds. Apparently, this is due to the low electrophilicity of DMCS and the difference in the electron structure of surface atoms of single-crystal silicon and silica.

It was established [11] that the activation energy of reactions of the electrophilic substitution of a silanol-group proton depends on the proton-acceptor properties of the attacking reagent; therefore it is possible to choose substituted silane capable of reacting with relatively inactive hydroxyl groups of surface silicon. This means that for surface reactions to occur one should either choose the reagent with $\sigma^*_i < \sigma_S^*$, or find the conditions of additional hydroxyl activation. The usage of triethoxysilane (TES) with (OC_2H_5) $\sigma^*_i = 1.366$ [4], the interaction of which with a silicon surface was studied within the temperature range of 300–673 K [9, 10], showed the thermostability of the product obtained at a temperature of 473K. This is indicative of its chemical bond with the surface, which can exist only during the reaction between silanol groups and ethoxy groups of sorbed silane according to the scheme:

$$(\equiv Si\text{-}OH)_m + (OC_2H_5)_3SiH \rightarrow (\equiv Si\text{-}O\text{-})_m\, Si(OC_2H_5)_{3-m}H + mC_2H_5OH$$

To determine the stoichiometry of surface reactions (numerical value of the coefficient m) one can also use the established correlation relation (3). By transforming it and replacing the (3 – m) alkoxy groups with $\sigma_i^*(OC_2H_5)=1,366$ [4] by the equivalent amount of OH-groups with $\sigma_i^*(OH) = 1.550$ [4], the observed shift of the frequencies of valence vibrations of the Si–H bond (around 4 cm–1 [10]) can be determined as

$$\Delta\nu(SiH) = a(3\text{-}m)OH[\sigma_i(OH) - \sigma_i(OC_2H_5)]$$

where $m = 3\text{-} \Delta\nu(SiH) / a[\sigma_i(OH)\text{-} \sigma_i(OC_2H_5)] = 3\text{-}4/19{,}21(1{,}550\text{-}1{,}366) = 1{,}86 \approx 2$.

Consequently, the structure of the surface compounds formed during the reaction with TES corresponds to the formula: $(\equiv Si\text{–}O)_2Si(OC_2H_5)H$. Using also Eq. (3) and substituting the experimental value $\nu(SiH)=2203cm^{-1}$ [10] for this surface compound in it, it is possible to estimate the inductive effect of silicon–oxygen groups on the surface of single-crystal silicon on the scale of inductive Taft constants:

$$\sigma_{S(Si)} = [(\nu_i\text{-} \nu_0)/a\text{-}\sigma_i(OC_2H_5)]/m = [(2203\text{-}2064)/19{,}21\text{-}1{,}366]/2 = 2{,}90.$$

A similar estimate of the inductive effect for the core of the binary semiconductor GaAs with the orientations (100) and (110) gave the calculated value σ_S(GaAs)=5.25. Processing of the surface of this semiconductor in the etching agent CH_3OH:Br (100:1) with subsequent washing in methanol and water reduced the value of σ_S(GaAs) to 5.10 (ν_i(Si–H) = 2162 cm_{-1}); this is caused by a significant decrease in the thickness of the oxide layer. The greater inductive effect of this matrix is due to the fact that its composition contains a more electronegative component, arsenic, and its oxides on the surface. Therefore there are grounds to assume that the hydroxyl groups on a real GaAs surface should have rather high proton-donor ability making it possible to use them in surface chemical reactions with relatively weak electrophilic reagents. This fact is of specific interest for the technology of surface nanostructures on GaAs.

Analysis of experimental results on determining the reactivity of hydroxyl groups of silica, silicon and arsenide–gallium matrices made it possible to assume that the hydroxyl groups on the real surface of a semiconductor matrix will possess an activity determined by the prehistory of the sample, namely, the thickness of the oxide layer formed after preliminary treatment. It is possible to determine this influence from the inductive effect and reveal the quantitative regularity of the reactivity of OH-groups of the surface of the crystal on the thickness of the oxide layer. Such experiments were performed on silicon samples (elements of FTIR) with a known oxide-layer thickness synthesized by the method of molecular layering (ML) [9, 10, 12]. The thickness of the layers was determined from ellipsometric measurements of parameters Δ and Ψ, and was calculated in the approximation of the Drude–Thornton single-layer model [13]. The obtained data made it possible to establish the quantitative relationship between the thickness of the oxide layer on the silicon matrix and its inductive effect, and, consequently, the activity of hydroxyl groups (Figure 1). The largest variations in σ^*_S were observed up to an oxide-layer thickness of $d_c \approx 3$ nm. The inductive effect of the silicon core with a larger oxide thickness was almost the same as for silica. As expected, the active chemosorption of DMCS was observed.

It should be taken into account that a silicon surface even freshly etched in aqueous solutions contains a thin oxide layer (according to our estimates of around 1nm) and the value $\sigma^*_{S(Si)} = 2.9$ calculated from the experimental values ν(Si–H)] characterizes exactly such a surface.

Estimation of the inductive constant performed according to the frequency of valence vibrations of Si-H groups formed directly in the process of silicon etching by hydrofluoric acid (ν (Si–H) = 2096 cm^{-1}) gave $\sigma^*_S = 1.63$. This value characterizes a silicon surface almost lacking an oxide layer and indicates the low activity of silanol groups bonded with the silicon core directly. The possible presence of fluorine groups on the surface can somewhat increase the value of $\sigma^*_{S(Si)}$, however extrapolation of the dependence $\sigma_S^* = f(d)$ to the value $\sigma^*_S = 1.63$ (Figure 1) gives a rather real thickness of the intrinsic residual oxide ($d \approx 1$nm) which coincides well with the ellipsometric estimate.

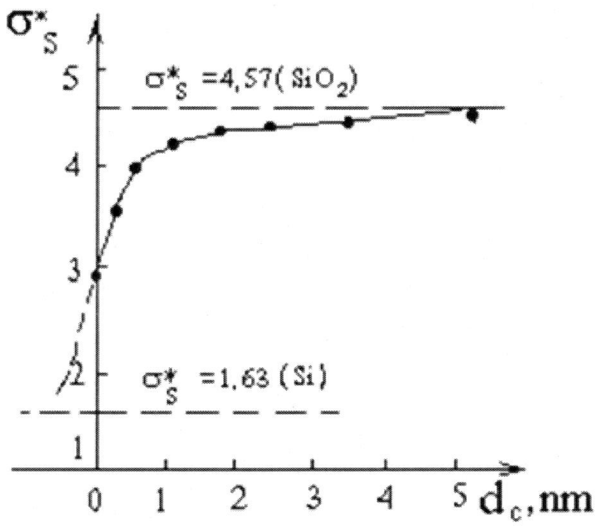

Figure 1. Effect of the thickness of the synthesized oxide layer d_c on the inductive constant σ_S of the silicon core.

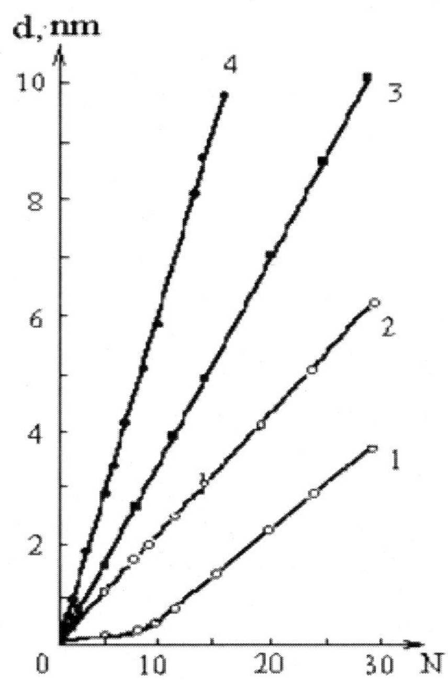

Figure 2. Growth dynamics of SiO_2 layers (1, 2), Ta_2O_5 (3) and Al_2O_3 (4) on silicon at a temperature of 523 K (1) in the absence and (2, 3, 4) presence of TEA on the number of treatment cycles of the silicon surface by metal halide and water vapors.

Thus, all real silicon matrices can be presented as a series of chemical compounds of the general formula $Si_m(SiO)_x(SiO_2)_yOH$, where a specific value corresponds to a specific composition (x, y), and at large y it corresponds to the final member of the series, i.e., hydroxylated silica $(SiO_2)_yOH$.

The values of the inductive constants make it possible not only to quantitatively estimate the effect of the core and the activity of OH-groups of solid compounds during heat treatment but also predict the direction of surface reactions with different reagents with the participation of these groups. To this end, it is necessary to compare the values of σ_i^* for different substituents with choice of the direction of the exchange of the group to a more electronegative one, i.e., with a larger inductive constant. For example, comparison of the inductive Taft constants for chlorine- and ethoxy groups (σ_i^* = 2.880 and 1.366, respectively [4]) and the value for freshly etched silicon (σ_S^*=2.9) shows that the reaction of hydroxyl groups of the surface of this matrix with compounds containing ethoxy groups is more preferable, since in this case there occurs an exchange to a substituent with a higher effective electronegativity. In the general case ($\sigma_S^*>\sigma_i^*$) is necessary as evidenced by the experiment. To study the surface reactions on objects with a small surface, e.g., single crystals, it is important to be able to determine their stoichiometry, when the use of direct chemical-analytical methods is quite complicated. The given example of solving this problem for a silicon matrix rather convincingly demonstrates the possibility of using inductive constants and the correlation analysis on the whole for determining the parameters of surface reactions.

The obtained values σ_S^* for a silicon matrix indicated the relatively weak proton-donor properties of silanol groups for a pure silicon surface and a surface containing a thin (<3.0 nm) oxide layer. This means that for the entire course of surface reactions one should either choose a reagent with $\sigma_\iota^*<\sigma_S^*$ or find conditions for the additional hydroxyl activation, which can be implemented either by increasing the temperature (see the table) or another activating effect, and also by using chemical reagents increasing the protonation of hydroxyls, i.e., proton acceptors. For example, the use of triethylamine (TEA) as an exchange catalyst made it possible to eliminate the latent period in the formation of the silicon-oxide layers caused by partial filling of the surface during its processing with $SiCl_4$ and H_20 vapors (Figure 2) [10]. The performed experiments showed (Figure 2) that TEA activated the growth of the SiO_2 layers and the increase in the thickness with the number of treatment cycles was in good agreement with the linear dependence $d = d_0N$ and the growth parameter d0 (the layer thickness per treatment cycle) was close to the calculated value for a monolayer.

Analogous experiments were performed during the synthesis of Al_2O_3 and Ta_20_5 nanolayers by alternating treatment of the silicon surface with Al_2Cl_6 or $TaCl_5$ and H_2O + TEA (Figure 2) vapors which also stabilized the dynamics of the growth of the oxide layer at different temperatures. As a result, the effect of the temperature of synthesis on the layer growth parameter was established (Figure 3) [14, 15]. This dependence makes it possible to make a judgment regarding the mechanism of layer formation and determine the conditions of the layer growth of nanostructures. For comparison Figure 3 shows the curves $d_0 = f(T_s)$ obtained without using TEA for SiO_2 and Al_2O_3 layers. In the studied temperature range the region of reduction d_0 was observed only for Ta_2O_5 at high temperatures; this is caused by the significant desorption of tantalum chloride.

The initial decrease in the parameter d_0 for all oxides indicates the polymolecular adsorption of reagents (condensation is possible for aluminum and tantalum halides) and their interaction in the adsorbed layer with the formation of a hydrated oxide. At T >423 K the chemosorption of halides has already a layerwise and distinguished activation character (Figure 3, curves 1a, 3a). It was possible to implement the layerwise growth of oxides in Si–SiO_2 and Si–Al_2O_3 systems without an activator only at T>550 K and T >500 K, respectively. Under these conditions their growth parameter approached the thickness of a monolayer for the given oxides. The use of TEA made it possible to eliminate the activation barrier that led not only to an increase in the chemosorption rate of halide, but also to a decrease in the temperature of layer growth by more than 100°C (Figure 3, curves 1b, 2, 3b).

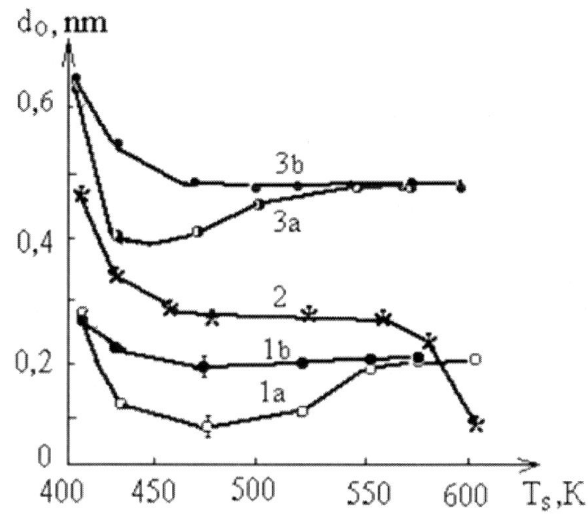

Figure 3. Effect of the temperature of synthesis on the growth parameter of SiO_2 (1a, 1b), Ta_2O_5 (2) and Al_2O_3 (3a, 3b) layers on silicon without an exchange activator (1a, 3a) and with an exchange activator (TEA) (1b, 2, 3b).

Conclusion

Precise chemical methods for the synthesis of low-dimensional objects, e.g., layers of nanometer thickness, on a solid surface are a component part of the intensively developing chemical nanotechnology of new materials with different end purposes. The developed approach and set of performed studies showed not only the possibility but also the prospects of using correlation analysis as the basis of the quantitative theory of surface reactions. The values of the inductive constants make it possible to quantitatively estimate the effect of the solid core and the state of its surface at different temperatures, predict the direction of surface reactions with different reagents and establish their stoichiometry, when the use of direct chemical-analytical methods is difficult. Such an approach can be the basis for further development of the chemical nanotechnology of low-dimensional systems using the method of molecular layering or atomic layer deposition.

REFERENCES

[1] V. A. Pal'm. *Bases of Quantitative Theory of Organic Reactions* (Khimiya, Leningrad, 1977) [in Russian].

[2] V. B. Aleskovskii, *Zh. Prikl. Khim.* **55**, (1982).725-730.

[3] V. B. Aleskovskii. *Course of Chemistry of Supramolecular Compounds* (SPb. Gos. Univ., St._Petersburg, 1996) [in Russian].

[4] *Chemist's Handbook* (Khimiya, Moscow, Leningrad, 1971), Vol.3 [in Russian].

[5] Yu. K. Ezhovskii, P. M. Vainshtein, and I. P. Gavrilina. *Chemical Physics Reports*. **2**. (1996).265-272.

[6] Yu. K. Ezhovskii, *Russ. Chem. Rev.* **2**, (2004), 195-204.

[7] Yu. K. Ezhovskii and A. V. Osipov. *Chemical Physics Reports*. **8**. (1996), 1187-1194.

[8] M. G. Voronkov. *Siloxane Bond* (Nauka, Novosibirsk, 1976) [in Russian].

[9] Yu. K. Ezhovskii. J. Surf. Inves. X-ray, *Synchrotron and Neutron Techniques*. 3. (2015). 462-467.

[10] Yu. K. Ezhovskii and P. M. Vainshtein, *Russ. J. Appl. Chem.* 7**1**, (2) (1998). 227-231.

[11] V. A. Tertykh, V. V. Pavlov, K. I. Tkachenko, and A. A. Chuiko, *Teor. Eksp. Khim.* **11**, (1975).174-177.

[12] S. V. Bukin and A. S. Shulakov, *J. Surf. Invest.: X-ray, Synchrotron Neutron Tech.* **1**, (2007), 67-70.

[13] R. Azzam and N. Bashara. *Ellipsometry and Polarized Light* (North-Holland, Amsterdam). (1977).

[14] Yu. K. Ezhovskii and A. I. Klusevich, *Inorg. Mater.* **39**, (10) (2003), 1062-1066.

[15] Yu. K. Ezhovskii, *Russ. J. Phys. Chem.* A **84**, (2010). 261-266.

In: Applied Aspects of Nano-Physics …
Editors: Kirill Levine and Andrey Syrkov
ISBN: 978-1-53614-708-7

Chapter 38

CARBOTHERMIC SYNTHESIS OF TITANIUM DIBORIDE: AN UPGRADE

*E. S. Gorlanov**
"Expert-Al" LLC, Saint Petersburg, Russia

ABSTRACT

The technology of carbothermic reduction of titanium and boron oxides in the TiO_2-B_2O_3-C reaction mixture at temperatures up to 1070°C is presented. The temperature and atmospheric regimes of TiO_2 low-temperature reduction are determined. In the Ti-B-O-C system, successive phase formation of $TiO_2 \rightarrow Ti_nO_{2n-1} \rightarrow TiBO_3 \rightarrow TiC_xO_{1-x} \rightarrow TiB_2$ at 1030 ÷ 1050°C was realized, which was recorded by X-ray phase analysis of the samples after synthesis. The conditions for the realization of low-temperature synthesis are established, which consist in observing the prescriptions and special regimes for preparing the reaction mixture, the conditions for heating and holding the samples in a controlled atmosphere. The mechanism of oxycarboboride phases formation at each of the stages of reduction of activated titanium oxide is presented for discussion. The principal possibility of low-temperature synthesis of TiB_2 in the composition of carbon-graphite products by free pitch technology in the conditions of existing kilns is shown.

Keywords: sol-gel technology, anatase-rutile transformation, phase formation, low-temperature synthesis, titanium oxycarbide, titanium diboride, dynamic vacuum, controlled atmosphere

* Corresponding Author's Email: gorlanove@yandex.ru.

Introduction

It is fairly believed that the development of the aerospace or nuclear industry and their military applications does not depend on the cost of innovative materials, and the strategic importance of these areas pays off any material costs. Therefore, it is these industries that have made it possible to improve the methods for synthesizing and compacting expensive powder materials, and to study and expand ideas about the useful properties of carbides, borides, borates and oxycarbides of transition metals. In the system Ti-B-C-O is formed a series of compounds of great importance for national economic development. In particular TiB_2, TiC and B_4C, possessing in a compact form high strength, thermal conductivity, resistance to wear, aggressive media and neutron irradiation, are used as refractories, wear-resistant and heat-resistant coatings and products, armor components, and as absorbing material for regulating rods of nuclear reactors [1]. Such compounds as TiO, $TiBO_3$ and TiC_xO_{1-x}, being thermally stable semiconductors and having good thermoelectric properties, are used as semiconductor and thermoelectric products in autonomous devices.

However, the complexity of high-temperature 1500 ÷ 2000°C production of powders of these materials causes their high cost and imposes serious restrictions on use in the national economic branches of general use. For example, for more than 50 years, the useful properties of TiO oxide semiconductors for thermoelectric devices [2] or titanium diboride TiB2 have been known as a metal-like conductor and refractory [3]. These materials have the potential for revolutionary innovations in metallurgy and in the utilization of "waste" heat from hot sources (surfaces of metallurgical aggregates, waste gases, etc.). However, the high cost (from 500 to 3000 $/kg of powder) and the lack of synthesis technologies in industrial volumes hamper their wide application. And this is an objective restriction, because such a high cost in comparison, for example, with carbon products cannot recoup the cost of new materials, even for the entire lifetime of metallurgical units.

In this connection, advanced scientific organizations, world manufacturers of refractories and metals in the last 10-15 years are conducting intensive research studies on the methods of synthesis of refractory materials based on oxides and borides of titanium with the aim of reducing their cost [4-9]. Some progress has been made, but commercial technologies for the production and use of these materials on a large scale have not been reported.

JUSTIFICATION AND PREPARATION OF CARBOTHERMIC SYNTHESIS

The present studies of the features and prospects of the refractory compounds synthesis in the Ti-B-O-C system have been undertaken to remove the economic limitations of the widespread use of titanium borides, oxides and oxycarbides in the metallurgical industry. It is assumed that the use of available and inexpensive starting components for their preparation in low-temperature conditions up to 1070°C using simple standard equipment will significantly reduce the cost of these powders to the level of commercial use. Such compounds are oxides of titanium and boron in the hydrated form of metatitanic ($H_2TiO_3 = TiO_2{\bullet}H_2O$) and boric ($H_3BO_3$ = $1/2B_2O_3{\bullet}3/2H_2O$) acids, which make it possible to use solution methods for mixing reagents to increase the reaction surface, in particular sol-gel technology.

But even in this case, the reactions of carbothermic reduction with the participation of three initial reagents at 1030 ÷ 1070°C are endothermic, which requires considerable energy spending for activating the synthesis of oxycarboborides.

$$TiO_2 + B_2O_3 + 5C \leftrightarrow TiB_2 + 5CO \quad \Delta G^0_{1300} = 210 \text{ kJ/mol} \tag{1}$$

According to the data of [10-12], reaction (1) at atmospheric pressure becomes thermodynamically possible ($\Delta G_R < 0$) only at 1327°C and at 1027°C, if the pressure in the system is reduced by 100 times. In addition, in the practice of carbothermic reduction, there is a need to increase energy spending by 20-25% relative to thermodynamically justified.

Overcoming these limitations is assumed using a known feature of titanium oxide - a change in the energy state during the phase transition of anatase to rutile a-TiO_2→r-TiO_2. During this period, the order of the crystal lattice is rearranged, the mobility of its constituent parts increases and, thus, the reactivity of titanium oxide is stimulated. But anatase-rutile transformation (ART) occurs in the temperature range 600 ÷ 800°C [13], whose energies are insufficient for the phase formation of the target products in the TiO_2-B_2O_3-C system. In this case, the reduction of stable inactive rutile r-TiO_2 by carbon becomes possible only in the interval 1350÷1550°C [11].

The most reliable way to stabilize the anatase is to introduce impurity elements into its lattice. In particular, doping of anatase with fluorine under certain conditions can significantly improve its stability [14, 15] up to the temperature and energy intervals, where possible carbothermic reaction synthesis of titanium oxycarbide in the system Ti-B-O-C. To implement them in the application to low-temperature conditions, the sol-gel preparation technology for titanium oxide in the initial form of metatitanic acid H_2TiO_3 was developed and implemented.

The process of its modification (doping) with fluorine-ion was carried out using hydrolysis according to the scheme:

$$TiO_2 \cdot xH_2O + NH_4OH + xHF \rightarrow TiO(OH)_{2-x}F_{x\downarrow} + xH_2O + NH_3 \quad (2)$$

which in the presence of hydrofluoric acid (fluoride ion source) and ammonium hydroxide (hydrolysis activator and pH regulator) proceeds to form a complex precipitate of metatitanic acid. With the heating of the amorphous doped mixture in an air atmosphere at 300 ÷ 400°C during the crystallization of the anatase from the sediment water is removed to obtain a doped anatase.

$$TiO(OH)_{2-x}F_x \rightarrow a\text{-}TiO_{2-x}F_x + H_2O\uparrow \quad (3)$$

To determine the ART intervals, the fluorine-modified titanium oxide was successively subjected to heating up to 1100°C in various atmospheres (Figure 1). Each of the obtained points in the graph reflects the amount of the rutile phase after heating and holding the unmodified (TiO_2) and doped ($TiO_{2-x}F_x$) samples at a fixed temperature for 15 minutes.

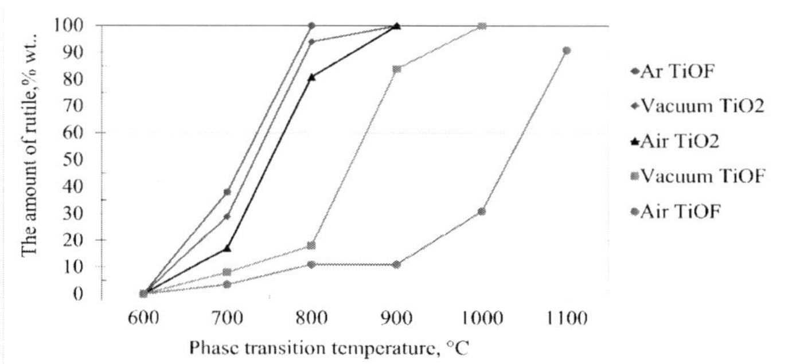

Figure 1. Dynamics of titanium oxide rutilization.

As follows from the obtained data, heating of amorphous titanium oxide in vacuum and especially in argon stimulates the ART transition. The increased reactivity of anatase a-$TiO_{2-x}F_x$ at 800 ÷ 1000°C for the longest period is maintained in the presence of oxygen. Therefore, the activation of a-$TiO_{2-x}F_x$ in the TiO_2-B_2O_3-C reaction mixture assumes its reduction with the formation of a homologous series of Ti_nO_{2n-1} oxides simultaneously with the ART process in the established temperature and atmospheric conditions. Simultaneously is a key definition of the reduction processes of modified titanium oxide. This means that not the final rutil phase r-TiO_2 interacts with the carbon, but the intermediate anatase with the crystal lattice in the process of reorganization:

$$a\text{-}TiO_{2-x}F_x + C \rightarrow Ti_nO_{2n-1} + xF\uparrow \ (x = 0 \div 1,\ n = 1 \div 4) \quad (4)$$

Thus, it is assumed that in the current temperature range 800÷1000°C, despite the insignificant endothermic barrier of reaction (4), the activated titanium oxide will be successively reduced by carbon to titanium carbide according to scheme

$$TiO_2 \rightarrow Ti_4O_7 \rightarrow Ti_3O_5 \rightarrow Ti_2O_3 \rightarrow TiO \rightarrow TiC$$

It is also important to take into account that the last stage of the TiO→TiC process is complicated by the formation of a continuous series of solid solutions. As a result, the resulting monoxide and carbide with a lack of oxygen and carbon can be considered as a solid solution of TiC-TiO or as titanium carbide, in the lattice of which a part of the carbon atoms is replaced by oxygen atoms TiC_xO_{1-x}. This means that this phase is in the process of rebuilding the crystal lattice and is also active for interaction in the reaction mixture TiO_2-B_2O_3-C.

Therefore, freshly formed non-stoichiometric oxycarbides TiC_xO_{1-x} with further heating and aging at 1030÷1070°C are the best starting phases for the interaction with carbon and boron components with subsequent formation of titanium diboride TiB_2.

$$TiC_xO_{1-x} + B_2O_3 + xC \rightarrow TiB_2 + yCO \quad (5)$$

Thus, it is intended to experimentally confirm and/or clarify the complete cycle of phase formation in the TiO_2-B_2O_3-C system.

EXPERIMENT

After doping the original amorphous titanium oxide, into its gel-like solution $TiO(OH)_{2-x}F_x$ the remaining components of the reaction mixture boric acid and carbon in the form of sucrose were added successively with constant stirring to a molar ratio of $TiO_2 : B_2O_3 : C$ from the stoichiometric ratio of 1: 1: 5 to 1: 5: 10. The resulting mixture was dried at room temperature, then at 80-90°C in a drying cabinet. The obtained agglomerated TiO_2-B_2O_3-C composite was heated in a water-cooled sealed cell with the possibility of adjusting the gaseous medium (Figure 2).

To maintain titanium oxide in the active state, heating up to 800 ÷ 1000°C was carried out in an air atmosphere. When the temperature reached 800 ÷ 1000°C, a dynamic vacuum was installed in the system to remove the evolved gases. After heating to the holding temperature of the system at 1030 ÷ 1070°C, a vacuum of about 1-3 kPa was established for 3-4 hours. In some cases, the completion of the experiment was carried out in an argon atmosphere.

After the end of the exposure at a fixed temperature and the completion of the low-temperature synthesis of the oxycarboboride phases, the reaction cell was cooled to room temperature followed by dismantling. Samples were extracted from the reaction zone, weighed and directed to X-ray phase analysis (radiation of CuKα).

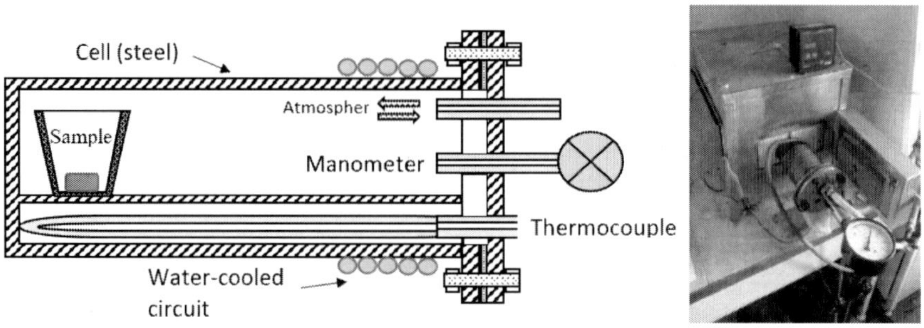

Figure 2. Reaction cell layout and assembly.

THE RESULTS OF THE EXPERIMENT AND THEIR DISCUSSION

First of all, it is necessary to verify the possibility of reducing the doped titanium oxide with carbon and determine the order and depth of phase formation when the reaction mixture is heated to 1050°C. For this purpose, compositions with a molar ratio of a-$TiO_{2-x}F_x$: C = 1:13 were prepared by a solution method. After drying, the black agglomerate was kept under argon atmosphere for 2 hours at 1050°C.

As follows from the presented results of the analysis of the reaction products a-$TiO_{2-x}F_x + C \rightarrow Ti_nO_{2n-1} + xF \uparrow$, the phase formation consistently developed from the anatase phase of a-$TiO_2 \rightarrow Ti_4O_7 \rightarrow Ti_3O_5 \rightarrow Ti_xO_{1-x}$ to non-stoichiometric titanium oxide. This means that eliminated 2 stages of phase transformations:

- *Rutilization of titanium oxide.* Phase formation develops directly from anatase modification, i.e., in the process of activating the crystal lattice and increasing the reactivity of TiO_2.
- *Formation of the Ti_2O_3 phase.*

Phase Ti2O3 is not detected, but two main peaks of the TixO1-x phase are clearly distinguished (Figure 3), i.e., energetically and kinetically the system is more profitable from Ti3O5 to proceed to the formation of nonstoichiometric titanium monoxide: 1/3Ti3O5 + C → TixO1-x + CO.

Thus, the doping of titanium oxide with fluorine allows the system to save energy on phase formation of the homologous series Ti_nO_{2n-1} and carry out the process at temperatures below 1070°C. This is confirmed by the data of the following series of experiments (Figure 4). At the soak temperature in a discharged atmosphere of up to

1000°C, only the first two stages of $TiO_2 \rightarrow Ti_4O_7 \rightarrow Ti_3O_5$ titanium oxide reduction develop with an initial transition to the release of a monoxide at 980°C: $Ti_3O_5 \rightarrow Ti_xO_{1-x}$.

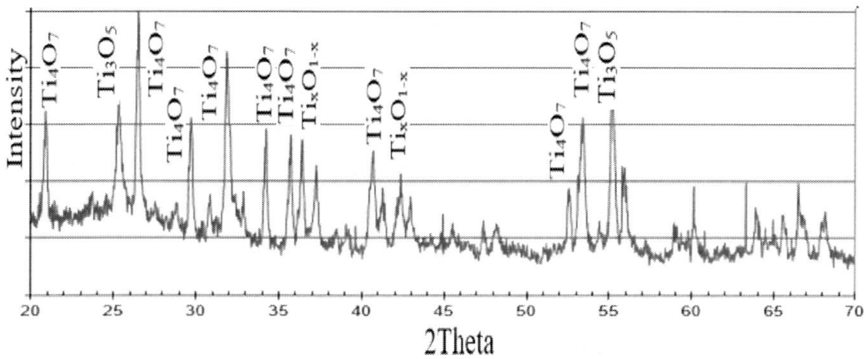

Figure 3. Results of X-ray phase analysis of $1TiO_2$: 13C samples after 2 hours at 1050°C in a discharged atmosphere.

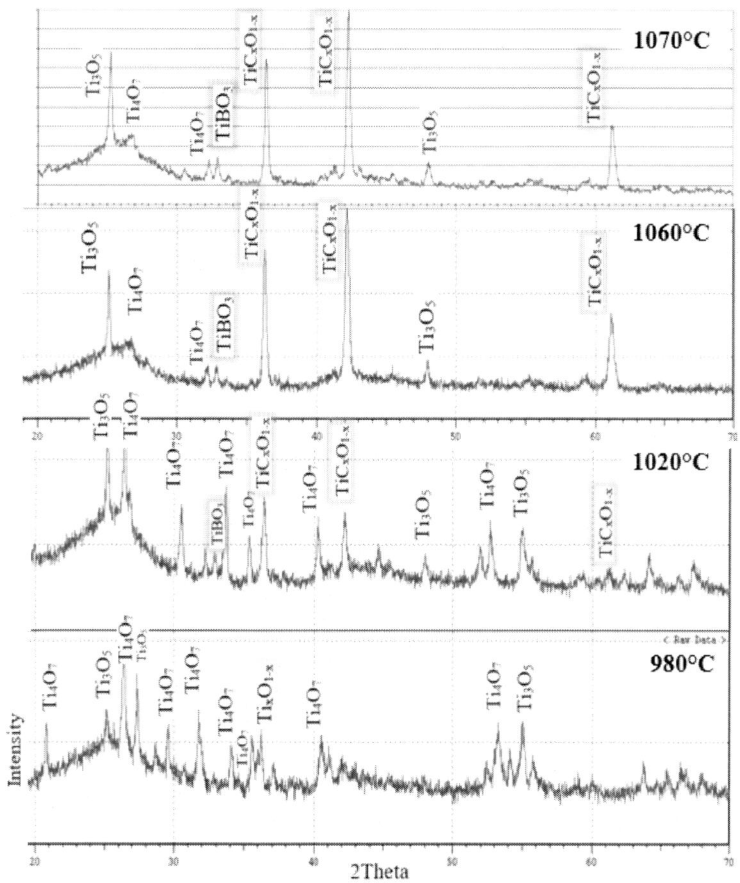

a) exposure 4 hours, vacuum

Figure 4. (Continued).

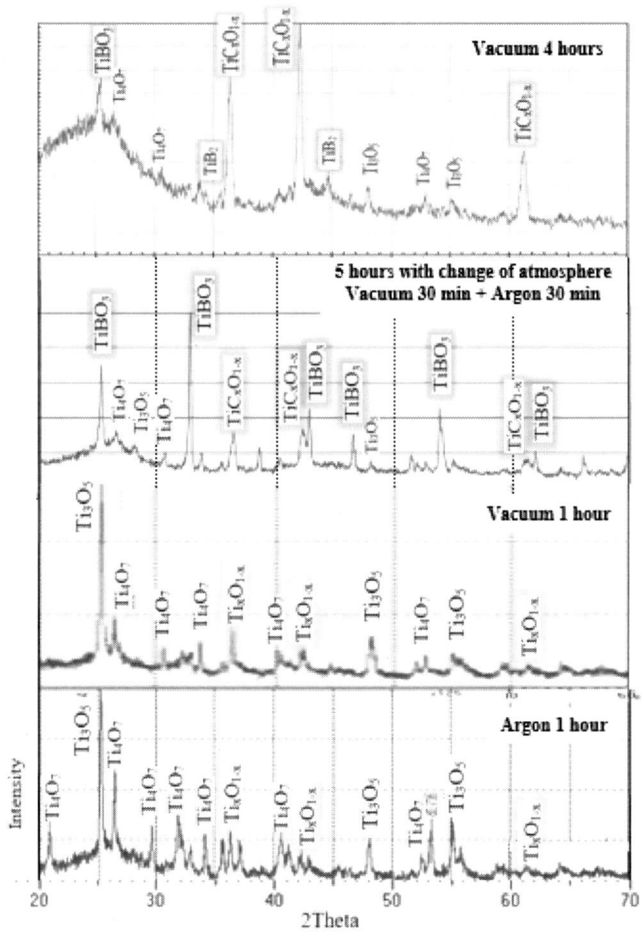

b) exposure at 1050°C

Figure 4. Results of X-ray phase analysis of samples $1TiO_2$-$2B_2O_3$-10C.

With increasing temperature, the process develops at a high rate. Holding for 1 hour in argon and vacuum atmospheres at 1050°C leads to an intensification of the formation of Ti_xO_{1-x} monoxide. With increasing duration of the exposure in the interval 800÷1070°C with the participation of Ti_xO_{1-x}, two competing processes develop - carbothermic reduction evolves to the non-stoichiometric phase of oxycarbide TiC_xO_{1-x} and in the contact with B_2O_3 the phase formation of yet another intermediate $TiBO_3$:

$$1/xTi_xO_{1-x} + (1/x + x)C \rightarrow TiC_xO_{1-x} + 1/Xco \quad (6)$$

$$1/xTi_xO_{1-x} + 1/2B_2O_3 \rightarrow TiBO_3 \quad (7)$$

In general, these two parallel processes can be represented as follows:

$$Ti_xO_{1-x} + 1/2B_2O_3 + yC \rightarrow TiBO_3 + TiC_xO_{1-x} + zCO \quad (8)$$

A similar dynamics of phase formation was observed by analyzing the products of synthesis at 1030°C in a combined atmosphere - successively for 3 hours holding the mixture in vacuo, replacing it with a 2-hour process in argon (Figure 5).

The limited activity of black carbon and an insufficient amount of boron oxide limit the phase formation with a mixture of simple $Ti_4O_7 \rightarrow Ti_3O_5$ and complex oxides of $TiBO_3$ and oxycarbides TiC_xO_{1-x} in accordance with reaction (8).

When the reaction stoichiometric mixture TiO_2-B_2O_3-C is formulated with sucrose, the activity of carbon together with the activated state of titanium oxide in the range 800 ÷ 1030°C is sufficient for the appearance of titanium borate as an intermediate phase (Figure 5b):

$$TiO_2 + 1/2B_2O_3 + 1/2C \rightarrow TiBO_3 + 1/2CO \quad (9),$$

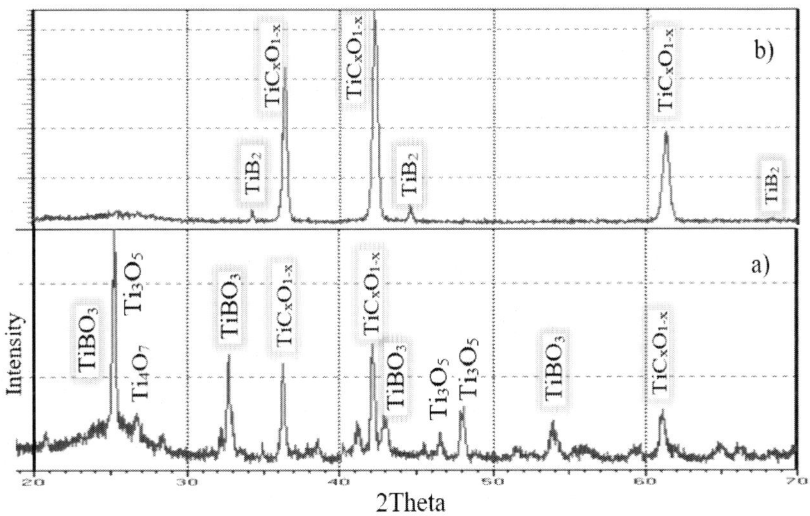

Figure 5. Results of XRD samples $1TiO_2$-$1,3B_2O_3$-5,4C after aging at 1030°C in a combined atmosphere (3 hours Vacuum + 2 hours Argon): a) carbon black, b) sucrose.

which, during aging at 1030°C, is completely consumed for interaction before the formation of boride and titanium oxycarbide:

$$2TiBO_3 + (5+2x)C \rightarrow TiB_2 + TiC_xO_{1-x} + (5+x)CO \quad 0 \leq x \leq 1 \quad (10)$$

Further recovery of titanium oxycarbide

$$TiC_xO_{1-x} + B_2O_3 + 2(2-x)C \rightarrow TiB_2 + (4-x)CO \quad (11)$$

did not develop due to the lack of sufficient boron oxide and carbon. With their shortage, the titanium oxycarbide with an admixture of titanium diboride TiB_2 (Figure 4b and 5b) was obtained in the experiments under consideration.

As a result, black agglomerates were obtained, a tangible difference which manifested itself in their strength. Samples obtained with a carbonaceous reductant in the form of soot are friable, samples with sucrose were more difficult to grind. For this reason, and in connection with the production of products with a small amount of amorphous phase (absence of "halo" by 2Θ to 30°), which distinguished them from previous synthesis results, the samples of synthesized powders were studied by computer microtomography (SkyScan 1272 microtomograph). It was found that the samples 5a consist of three phases (Figure 6), which, according to X-ray diffraction data and X-ray density distribution, obviously correspond to Ti_3O_5, TiC_xO_{1-x} and $TiBO_3$. Samples 5b were identified as TiB_2 and TiC_xO_{1-x} (Figure 6).

Thus, according to the results of the studies, the following sequence of carbothermic reduction of titanium oxide is detected:

$$TiO_2 \rightarrow Ti_nO_{2n-1} \rightarrow TiBO_3 \rightarrow TiC_xO_{1-x} + TiB_2 \rightarrow TiB_2 \quad (12)$$

With the correct composition of the initial reaction mixture and a sufficient number of initial components, the final phase of the synthesis is titanium diboride TiB_2.

Taking into account the established features of the phase formation in the TiO_2-B_2O_3-C system and the dependence of the yield of the final product on the initial composition of the mixture, atmosphere, and the temperature regime of the synthesis, a series of experiments with an increased excess of boron oxide was carried out. In Figure 7 presents the results of these studies.

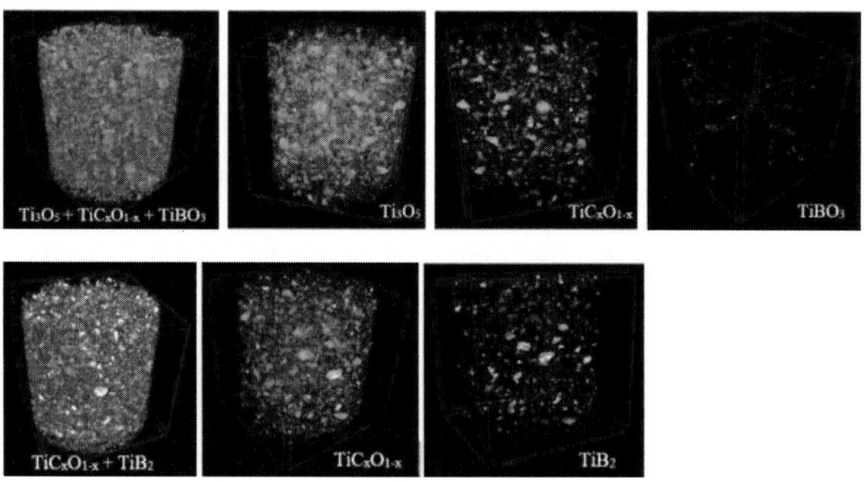

Figure 6. Microtomography of samples 5a (upper) and 5b (lower).

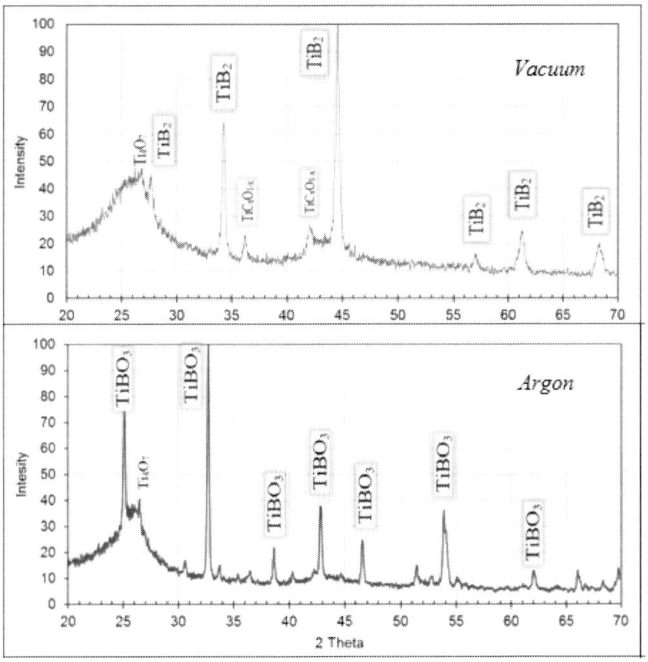

Figure 7. The results of XRD samples of $1TiO_2$-$5{,}4B_2O_3$-10C after holding for 4 hours at 1050°C.

Synthesis of $1TiO_2$-$5{,}4B_2O_3$-10C mixture in an argon atmosphere at 1050°C produced titanium diboride-titanium borate $TiBO_3$ with traces of titanium oxides, which is due to the presence of oxygen in an inert gas:

$$TiB_2 + 9/4O_2 = TiBO_3 + 1/2B_2O_3\uparrow \quad (13)$$

The experiment in vacuum produced titanium diboride TiB_2 with traces of unreduced titanium oxycarbide:

$$TiC_xO_{1-x} + B_2O_3 + 2(2\text{-}x)C \rightarrow TiB_2 + (4\text{-}x)CO \quad (14)$$

At the next stage, the preparation and calcination conditions of the graphite/anthracite carbon mixture (G/A = 50/50) with titanium and boron oxides were simulated. A composition (G/A) : ($1TiO_2$-$5{,}5B_2O_3$-8C) = 46 : 54 (wt%) was prepared by mixing the dry carbon and gelled oxide components, followed by pressing at 200 kgf/cm^2, drying the tableted "green" preforms at 60 -80°C and pre-firing in air at 450°C under a layer of petroleum coke.

Pre-prepared samples with a diameter of 26 mm and a height of 8-10 mm were placed in a cell (Figure 2) under a layer of coke, followed by heating and holding in a moderate vacuum of about 2 kPa at 1050°C for 4 hours. The results of XRD products of low-temperature processing are shown in Figure 8.

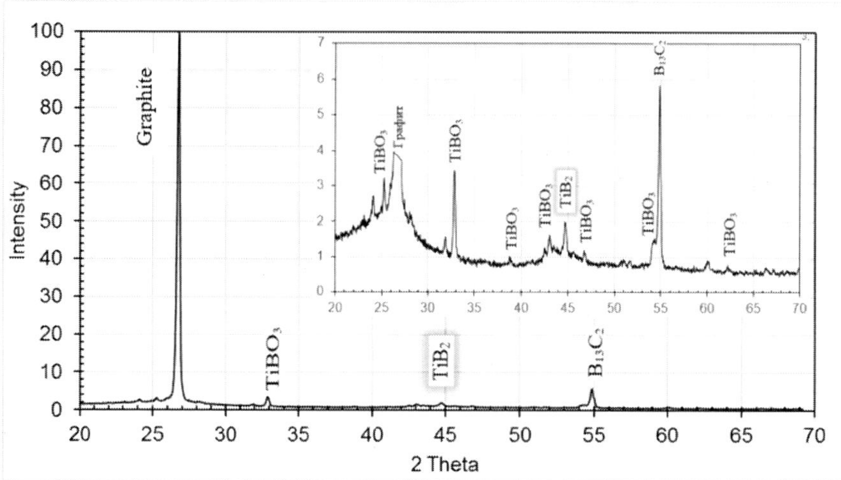

Figure 8. The results of XRD samples (G/A) : ($1TiO_2$-$5,5B_2O_3$-8C) = 46 : 54 after holding in vacuum at 1050°C for 4 hours.

On the background of the main graphite component, only the main peaks of TiB_2, $TiBO_3$ and $B_{13}C_2$ with very low intensity can be identified, which indicates that these compounds are not very high in the final product of low-temperature processing. The method and conditions for the preparation of the initial mixture suggest that the bulk of them are in the intergranular space, that is, the binder of graphite grains of calcined samples. For more information about their presence, the graphite peak is partially removed (the field in the x-ray angle).

The obtained data testify to the principle possibility of synthesis TiB_2 under conditions close to the conditions for calcination of cathode carbon-graphite blocks. The presence of oxygen in the CO/CO_2 gas mixture leads to oxidation of titanium diboride in the composition of carbon samples to titanium borate $TiBO_3$. Therefore, further studies will be aimed at increasing the concentration of titanium diboride in the composition of carbon products and reducing the effect of the gaseous atmosphere of the calcination cell on the final composition of the composition Carbon – TiB_2.

Conclusion

Doping of amorphous titanium oxide with fluorine allows in the system TiO_2–C to carry out a full cycle of phase formation in a series of Ti_nO_{2n-1} at temperatures below 1000°C.

The successive phase formation of $TiO_2 \rightarrow Ti_nO_{2n-1} \rightarrow TiBO_3 \rightarrow TiC_xO_{1-x} \rightarrow TiB_2$ is realized in the TiO_2-B_2O_3-C system under low-temperature synthesis conditions at 1030 ÷ 1050°C.

Depending on the initial composition of the reaction mixture and the final synthesis conditions, one can produce individual products – $TiBO_3$, TiC_xO_{1-x} and TiB_2.

The principal possibility of low-temperature synthesis of TiB_2 in the composition of carbon graphite products under conditions of low partial pressure of oxygen in the gas mixture of kilns is shown.

The reaction mixture TiO_2-B_2O_3-C in the form of a gel can be used as a binder for making carbon or other refractory products (pitch-free technology).

REFERENCES

[1] Properties, production and application of refractory compounds. Ref. ed. Ed. Kosolapova T. Ya. – M.: *Metallurgy*, 1986, 928 p. (In Russ.).

[2] Ioffe, A. F. *Semiconductor thermoelements*. A.F. Ioffe; Academy of Sciences of the USSR, Institute of Semiconductors. M.; L.: Publishing House of the USSR Academy of Sciences, 1960. - 188 p. (In Russian).

[3] Pat. 3028324 US. Int. Cl. 204-67. Producing or refining aluminum. Ransley C. E.; *The British Aluminium Company*. – Appl. No. 660994; Filed: May 23, 1957; Date of Patent: Apr. 3, 1962. – 34 p.

[4] Axelbaum, R. L. Wet chemistry and combustion synthesis of nanoparticles of TiB_2/R.L. Axelbaum et al. *Nano Structured Materials.* 1993. No.2. P.139-147.

[5] Krishnarao, R. V. Studies on the formation of TiB_2 through carbothermal reduction of TiO_2 and B_2O_3/R.V. Krishnarao, J. Subrahmanyam. *Materials Science and Engineering* A362. 2003. P.145-151.

[6] Chen, L. A facile one-step route to nanocrystalline TiB_2 powders/L. Chen et al. *Materials Research Bulletin.* – 2004. – No.39. – P.609–613.

[7] Ziemnicka-Sylwester, M. TiB_2-Based Composites for Ultra-High-Temperature Devices, Fabricated by SHS, Combining Strong and Weak Exothermic Reactions/M. Ziemnicka-Sylwester. *Materials.* 2013. No. 6. – P.1903-1919. - doi:10.3390/ma6051903.

[8] Pat. 2498880 The Russian Federation. Int. B22F, C01B, C04B. *Method for obtaining titanium diboride powder for the material of the wettable cathode of an aluminum electrolyzer*. Ivanov V. V. et al.; Applicant and patent holder of the Federal State Educational Establishment of Higher Professional Education "Siberian Federal University". 2012134603/02; claimed. 13.08.2012; publ. 11/20/2013, Bl. № 32. - 2 p. (In Russian).

[9] Blokhina, I. A. *Carbothermic synthesis and oxidation of* TiB_2 *powders: special. 05.16.06 - powder metallurgy and composite materials*. Diss. to the soisk. uch. Art. Cand. tech. sciences. Krasnoyarsk, 2015. - 122 p. (In Russian).

[10] Bagdavadze, J. Thermodynamic analysis of the Ti-O-C system/J. Bagdavadze, Z. Tsiskaridze and K. Ukleba. *Eur. Chem. Bull.* 2015. No. 4(3). P.128-129.

[11] Ma, Ai-qiong. Reactionary mass transfer mechanism of TiB_2 synthesized by carbothermal reduction method/Ai-qiong Ma, Ming-xue Jiang, Zhihong WU//*The Chinese Journal of Nonferrous Metals.* 2013. Vol.23. № 6. P. 1605-1610.

[12] Ma, A.-q. Reactionary mass transfer mechanism of TiB_2 synthesized by carbothermal reduction method/A.-q. Ma, M.-x. Jiang, Zh. Wu// *The Chinese Journal of Nonferrous Metals*. 2013. Vol.23. № 6. P. 1605-1610.

[13] Hanaor, D. A. H. Review of the anatase to rutile phase transformation/D. A. H. Hanaor and Ch. C. Sorrell. *J. Mater. Sci.* 2011. Vol.46. P. 855-874

[14] Yu, J. C. Effects of F^- Doping on the Photocatalytic Activity and Microstructures of Nanocrystalline TiO_2 Powders, J. C. Yu, et al. *Chem. Mater.* 2002. Vol. 14. P. 3808-3816.

[15] Sedneva, T. A. Dependence of phase transitions and photocatalytic activity of nano-sized titanium dioxide on doping with fluoride ions. T.A. Sedneva et al. *Perspective Materials.* - 2007. - №6. - P. 49-55. (In Russ.).

In: Applied Aspects of Nano-Physics …
Editors: Kirill Levine and Andrey Syrkov

ISBN: 978-1-53614-708-7

Chapter 39

PREPARATION OF NANOPARTICLES OF GERMANIUM VIA THE CATALYTIC METHOD

***Alena V. Kadomceva*[1,*] *and Anatoly M. Obedkov*[2]**
[1]Nizhny Novgorod State Medical Academy, Nizhny Novgorod, Russia
[2]Razuvaev Institute of Organometallic Chemistry, Nizhny Novgorod, Russia

ABSTRACT

A method was developed for reduction of germanium tetrachloride in the presence of a catalyst based. The method makes it possible to reduce the process temperature and diminish the number of stages in production of germanium. The kinetic characteristics of the catalytic reduction of germanium tetrachloride with hydrogen were determined. We have studied the kinetics of the catalytic reduction of germanium tetrachloride with hydrogen in the temperature range in the presence of the composites as catalysts and determined the reaction order and activation energy for the catalytic reduction of germanium tetrachloride with hydrogen.

Keywords: germanium tetrachloride, germanium, nanoparticles, reduction, catalyst, germanium nanoparticles, the zero order of the reaction, diffusion, germanium recovery, conversion

* Corresponding Author's Email: kadomtseva@nizhgma.ru.

INTRODUCTION

In recent years, the development of bulk nanostructured metallic materials has become one of the most relevant areas of modern nanomaterial science. The creation of nanostructures in metals and alloys opens the way for obtaining unusual properties that are very attractive for innovative applications. Particular attention is paid to catalytic processes. In addition, now there is a transition from laboratory methods to the creation of pilot - industrial technologies.

The study of nanoparticles and nanomaterials based on germanium, as well as its compounds, began relatively recently. Germanium compounds are used to produce spherical nanoparticles (Ge embedded in Al or In), nanotubes (epitaxial heterostructure $Ge_{0.4}$ $Si_{0.6}$), nanocrystalline magnetosensitive films (Fe_xGe_{1-x}). As components of composite nanomaterials, GeO_2 is actively used. So, nanowires and nanowires were made on the basis of the GeO_2-Fe_2O_3 composite, and nanowires and promising electrical and magnetic materials were made from the GeO_2-SiO_2 composite.

MATERIALS AND METHODS

The method of obtaining germanium nanoparticles is very complex and time-consuming. In this paper, the possibility of obtaining germanium nanoparticles by the catalytic method is considered.

A method of extracting germanium from a germanium-containing material is known [1]. The invention relates to the field of metallurgy of dispersed metals, mainly to the processing of germanium-containing raw materials, in particular combustion products of germanium-containing coals, to produce enriched germanium sublimes by an electric melting process. The disadvantage of the process is a large number of initial reagents and process steps, high contamination level of the final product, large reaction temperatures, high energy efficiency of the germanium production process, a decrease in furnace productivity and an increase in electricity and electrode consumption by 2 to 2.5 times is observed.

A method for extracting germanium is known [2], based on mixing of the initial charge containing a germanium-containing material, a sulfidizer and a reducing agent, agglomeration with a moistening and subsequent electric melting with a periodic charge of the charge on the previously induced melt. In the process of electric melting of germanium-containing raw materials, an insignificant amount of germanium is formed (2-6%), which is the main disadvantage of the method.

The considered methods of obtaining germanium have a number of significant drawbacks: practically all methods are not selective, as a result a rather large amount of by-products is formed that require a complicated and costly purification technology.

Thus, the search for new methods of obtaining germanium by the reduction of germanium-containing compounds is of great interest from an applied and fundamental point of view.

A new method for the catalytic reduction of germanium tetrachloride by hydrogen allows the production of germanium nanopowder with a size up to 140 nm as a product.

Results and Discussion

According to the gas chromatographic analysis, the dependence of the change in the concentration of germanium tetrachloride on temperature was obtained.

It was found that the reduction of germanium tetrachloride by hydrogen on various catalysts proceeds over a wide range of temperatures, with different conversions for germanium tetrachloride [3].

In order to establish the effect of diffusion of germanium tetrachloride on the surface of the catalyst, hydrogen reduction was carried out at different flow rates. It was found that the concentration of germanium tetrachloride at different flow rates remains practically constant, and a multiple change in the height of the catalyst bed (from 0.05 to 0.20 m) does not lead to a change in the quantitative composition of the gas mixture, which indicates the absence of diffusion limitations shown in Figure 1.

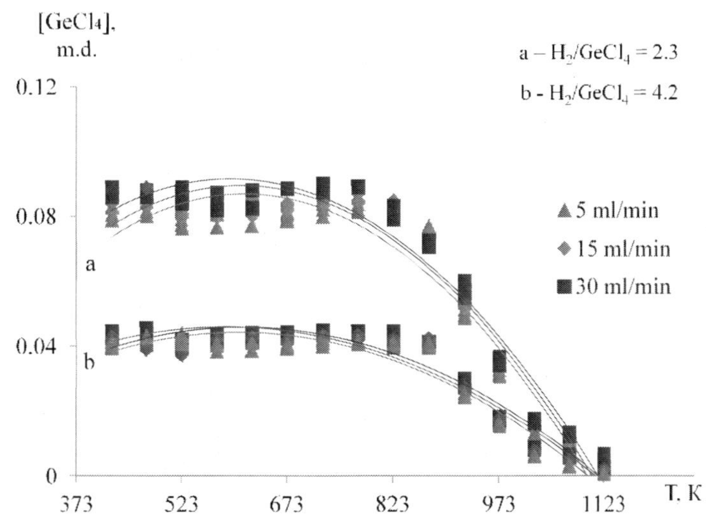

Figure 1. Dependence of the concentration of germanium tetrachloridefrom the temperature at different hydrogen flow rates.

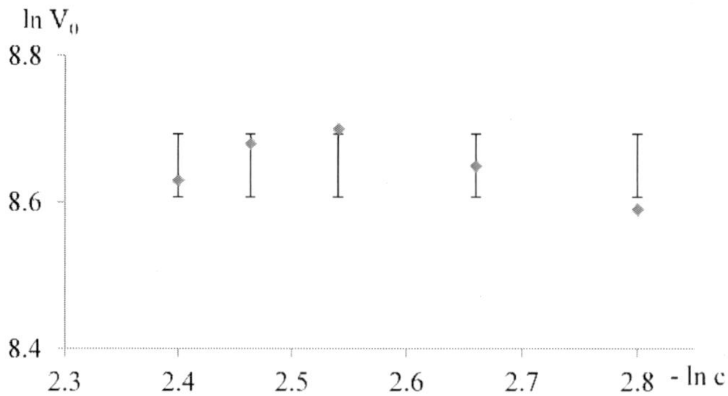

Figure 2. The logarithmic dependence of the reaction rate on the concentration.

The reduction of germanium tetrachloride by hydrogen is described by zero order equations, which indicates a high adsorption of the reactants on the catalyst, so the reaction rate is practically independent of the concentration of reagents in the vapor, since the diffusion rate of the reacting component to the active center of the catalyst is greater than the release rate of the active center after the reaction proceeds and the concentration decreases linearly with time shown in Figure 2.

$$\ln V_0 = (9.03 \pm 2.71) - (0.14 \pm 0.04) \ln C,$$

where V_0 – chemical reaction rate, $mol \cdot l^{-1} \cdot c^{-1}$; C – the current concentration of germanium tetrachloride, $mol \cdot l^{-1}$.

As a result of processing the obtained data by the method of least squares, equations were obtained for determining the rate constant for the catalytic reduction of germanium tetrachloride by hydrogen in the presence of various catalysts [4].

Quantitative analysis of the formed substances was carried out using a gas chromatographic method using a thermal conductivity detector using a Tsvet-800 gas chromatograph with a vacuum sample inlet system. To separate the reaction products, a packed column packed with N-AW-HMDS (0.16 to 0.20 mm) chromatography with 15% of E-301 liquid phase applied at 373 K for 6 minutes, a carrier gas of hydrogen (99.9999%) [5].

Processing of the results was carried out in the program “Color-Analytic” using relative graduation.

Conclusion

The reduction of germanium tetrachloride with hydrogen in the presence of a catalyst proceeds with 98.9% conversion by germanium tetrachloride at a temperature of 973K.

The characterization of the extracted germanium particles was carried out using atomic force microscopy, in addition, the size of the obtained germanium nanopowder was determined using a Shimadzu SALD-2300 laser analyzer. Shimadzu AA-7000 atomic absorption spectrophotometer was used to determine the possibility of contamination of the germanium nanopowder obtained. It was found that the catalyst practically does not increase the content of foreign matter in the germanium nanopowder composition.

The reliability of scientific results is justified by the use of modern analytical methods with high detection limits.

REFERENCES

[1] Patent RU № 2375481.

[2] Patent RU № 2385355.

[3] Kadomtseva, A.V., Vorotyntsev, A.V., Vorotyntsev, V.M., et al., *Russ. J. Appl. Chem,* 88 (4), (2015) 595–602.

[4] Kadomtseva, A.V., Vorotyntsev, A.V., Zelentsov, S.V., Vorotyntsev, V.M., et al., *Russ. Chem. Bull.,* 64 (4), (2015) 759–765.

[5] Kadomtseva, A.V., Ob"edkov A. M., Semenov N. M., Kaverin B. S., and Gusev S. A., Russ. *J. Appl. Chem,* 89(11), (2016) 1795−1803.

In: Applied Aspects of Nano-Physics …
Editors: Kirill Levine and Andrey Syrkov
ISBN: 978-1-53614-708-7

Chapter 40

ATOMIC FORCE MICROSCOPY OF NANOCOMPOSITES BASED ON ZINC OXIDE WITH DIFFERENT ADDITIVES

Evgeniya V. Maraeva*[1,2,*]*, Vyacheslav A. Moshnikov*[1,2]*,
Nadezhda D. Yakusheva*[2]*, Igor A. Pronin*[2]
***and Igor A. Averin*[2]**
[1]Saint Petersburg Electrotechnical University "LETI,"
Russian Federation
[2]Penza State University, Russian Federation

ABSTRACT

This chapter is concerned with the investigation of nanomaterials based on ZnO with different additives (Fe, Cu). A correlation between technological modes of production (annealing temperature, composition) and the values of the fractal dimension of the samples is discussed below.

Keywords: porous nanoparticles, atomic force microscopy, fractal dimension, zinc oxide

[*] Corresponding Author's Email: jenvmar@mail.ru.

Introduction

Currently, the nanocomposites based on metal oxides are widely used for creation gas sensors with a percolation structure [1-6]. The devices based on such structures allow one to reach extremely high values of gas-sensitive [7-10]. One of the most important ways to study such structures is fractal analysis based on data getting with the use of atomic force microscopy (AFM). There are different approaches to estimate the fractal dimension of the nanostructures, such as method of cube counting, triangulation method, power spectrum method etc. However, all of these methods depend on many factors (for example the size of scan area). All of them can be implemented in the graphic program "Gwyddion" for scanning probe microscopy (SPM) data analysis.

Materials and Methods

In this work the study of surface morphology of nanocomposites based on zinc oxide were carried out by means of scanning probe microscopy with atomic - force microscope NTEGRA (NT-MDT, Russia) with the use of silicon probes NSG with a radius of 10 nm in tapping mode. Evaluation of the fractal dimension was carried out in a graphics program for SPM data analysis «Gwyddion». The methods of cube counting and triangulation method based on the analysis of AFM images were used.

The main purpose of this work was to study the peculiarities of the surface relief in the nanocomposites based on zinc oxide and to determine the influence of annealing temperature and composition of investigated materials on the value of their fractal dimension. The stated purpose implies the following objectives:

1. To study the nanocomposites based on zinc oxide with different additives (Fe, Cu) with the use of scanning probe microscopy;
2. To identify the relationship between technological modes of production (composition, annealing temperature) and the values of fractal dimension of the samples.

Results and Discussion

For research purposes, a set of samples based on zinc oxide with additives of Fe and Cu, heated at different temperatures (420 K, 870 K) was obtained [10-12].

With the use of atomic force microscopy method it was found that all the samples were characterized by the presence of close-packed grain system. For example, in the

Figures 1, 2 shows the AFM image of the sample surface composition of the ZnO (no additives), annealed at T=150 °C. Images correspond to different sizes of the scan area.

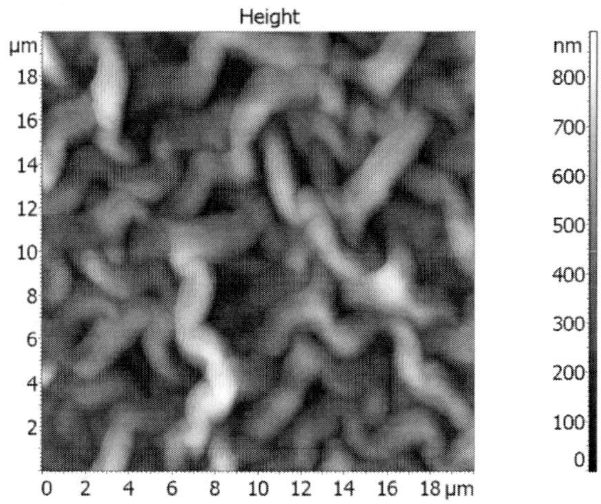

Figure 1. AFM image of the sample surface composition of the ZnO annealed at T=150 C, scan size 20*20 μm.

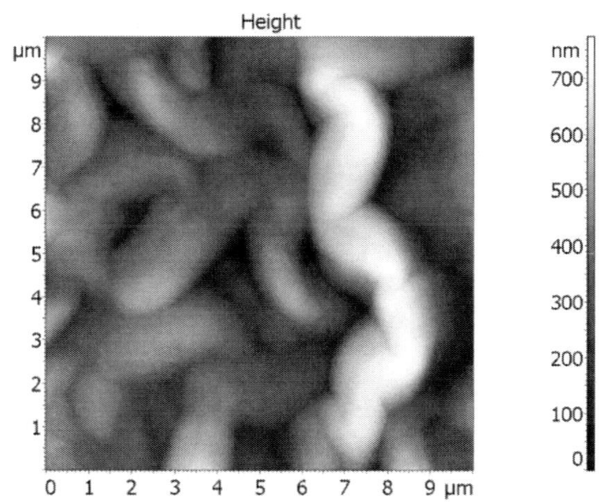

Figure 2. AFM image of the sample surface composition of the ZnO annealed at T=150 °C, scan size 10*10 μm.

From the morphological point of view the layers without additives of Cu and Fe were uniform despite of the processing conditions, in all parts of the surface the same relief was observed. The range of heights at different areas of a surface was 700 – 1200 nm. For example, Figure 3 shows the AFM-image of surface sample of ZnO annealed at T=600 °C (the dimensions of the scan area 50*50 μm) and the cross section of the sample.

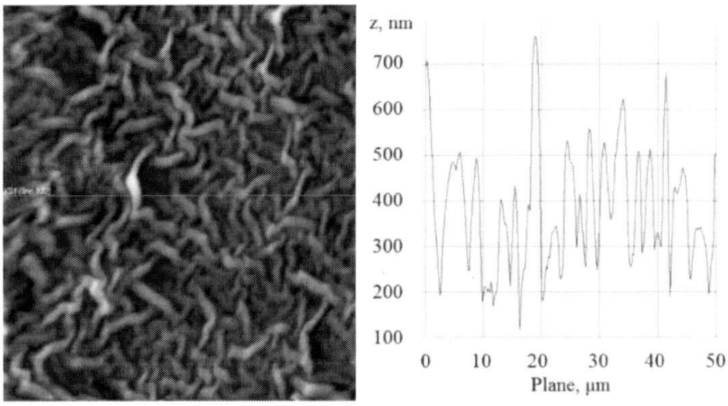

Figure 3. AFM image of the surface and the cross section of the ZnO sample.

In some cases (the samples with Cu and Fe additives) the surface contained inclusions that represented the core with radiating channels. For example, Figure 4 shows the AFM-image of surface sample of ZnO (Fe) after heat treatment at T=150 °C.

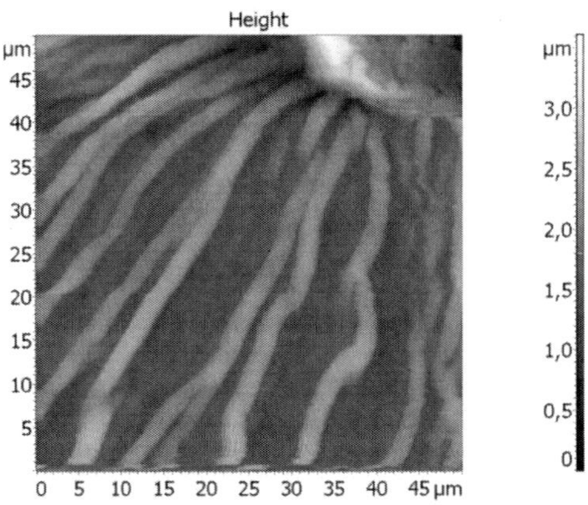

Figure 4. AFM-image of area 1 of the surface of the sample ZnO (Fe), annealed at T=150 °C, scan size of 50*50 μm.

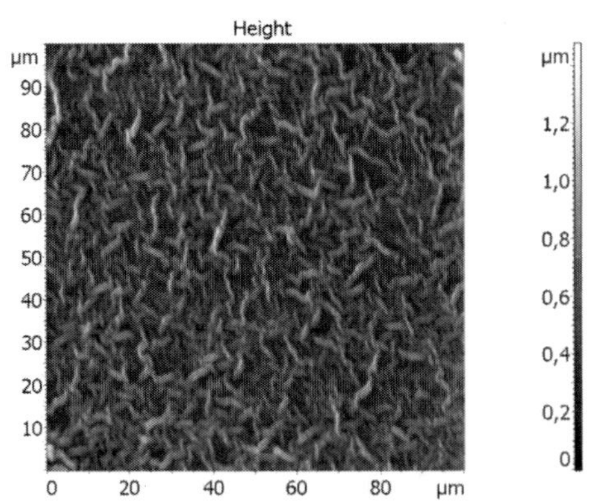

Figure 5. AFM image of the surface of the sample ZnO (Fe), annealed at T=150 °C, scan size 100*100 μm.

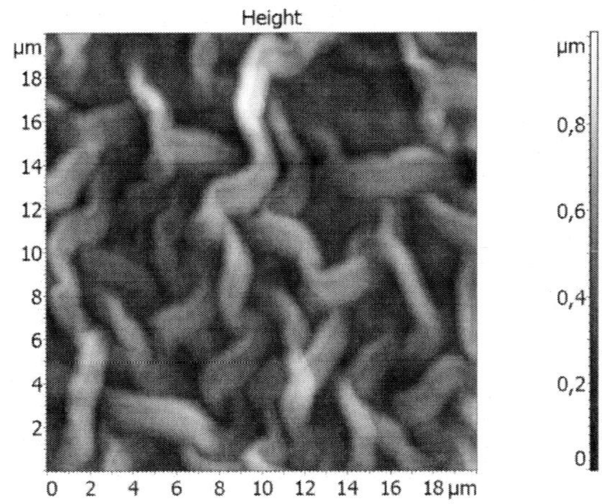

Figure 6. AFM image of the surface of the sample ZnO (Fe), annealed at T=150 °C, scan size 20*20 μm.

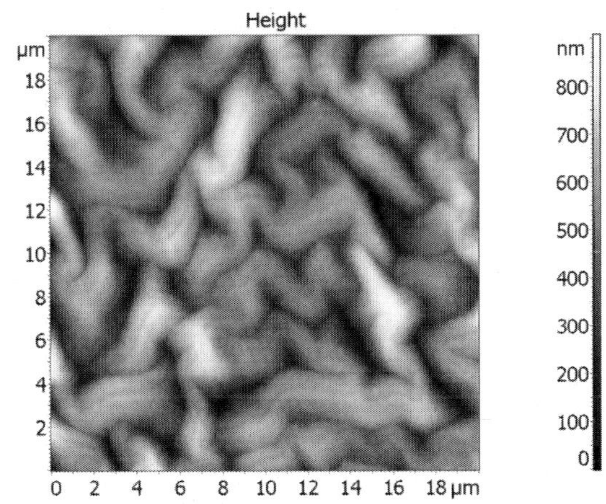

Figure 7. AFM surface image of the sample ZnO (Cu), annealed at T=150 °C, scan size 20*20 μm.

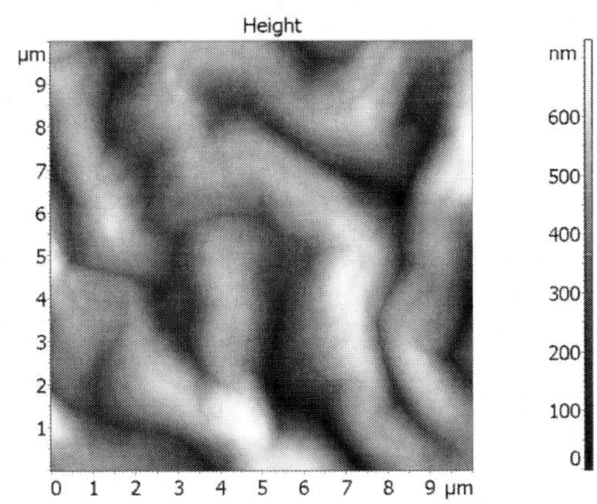

Figure 8. AFM image of the surface of the sample ZnO (Cu), annealed at T = 150 °C, scan size 10*10 μm.

The surface areas free from impurities were characterized by typical samples relief (the Figures 5 – 8).

For all the patterns the calculation of fractal dimension values was carried out with the use of the program "Gwyddion." The results are shown in the Table 1.

Table 1. The results of fractal analyses

Sample number	1a	2a	3a	1b	2b	3b
The composition of the sample	ZnO	ZnO (Fe)	ZnO (Cu)	ZnO	ZnO (Fe)	ZnO (Cu)
Annealing temperature, K	870	870	870	420	420	420
Fractal dimension (triangulation method)	2,54	2,60	2,55	2,58	2,58	2,33
Fractal dimension (method of cube counting)	2,47	2,52	2,46	2,47	2,48	2,36

It can be seen from the Table 1 that for ZnO (Cu) pattern which was heat-treated at T = 420 K the surface structure was characterized by the lowest value of the fractal dimension.

Thus, the atomic force microscopy data analysis showed that the presence of alloying elements (Fe, Cu) primarily affects the distribution of grains on the surface of the zinc oxide nanocomposites. The variation of the heat treatment regimes in the range from 420 K to 870 K slightly affects the surface morphology of the samples. The lowest value of the fractal dimension characterizes the surface of ZnO (Cu) nanocomposites, annealed at T = 420 K. For other types of the samples the fractal dimension varies slightly depending on the conditions of synthesis and depends mainly on the method of calculation.

ACKNOWLEDGMENTS

The reported study was supported by the Ministry of Education and Science of the Russian Federation within the framework of the project № 16.897.2017/PCh.

REFERENCES

[1] Pronin, I.A., Donkova, B.V. Dimitrov, D.Tz. etc. Relationship between the photocatalytic and photoluminescence properties of zinc oxide doped with copper and manganese. *Semiconductors*, 2014. Vol. 48, № 7. PP. 842–847.

[2] Abrashova, E.V. Kononova, I.E. Moshnikov, V.A. Metal oxide SnO_2 - ZnO - SiO_2 films prepared by sol-gel. *Smart Nanocomposites*, 2013. Vol, 4, № 2. PP. 1-7.

[3] Kalinina, M.V. Moshnikov, V.A. Tikhonov, P.A., etc. Temperature dependence of the resistivity for metal-oxide semiconductors based on tin dioxide. *Glass Physics and Chemistry*, 2003. Vol. 29, №. 4. PP. 422-427.

[4] Maraeva, E.V. Istomina, M.S. Moshnikov, V.A. etc. Study of porous sol-gel nanocomposites based on silicon dioxide and tin dioxide modified by fullerenol $C_{60}(OH)_n$ (n = 22–24). *Journal of Physics: Conference Series*, 2016. Vol. 690. P. 012031.

[5] Averin, I.A. Pronin, I.A. Kaneva, N.V. etc. Photocatalytic oxidation of pharmaceuticals on thin nanostructured zinc oxide films. *Kinetics and Catalysis*, 2014. Vol. 55, No. 2. PP. 167–171.

[6] Averin, I.A. Pronin, I.A. Bozhinova, A.S. etc. The thermovoltaic effect in zinc oxide inhomogeneously doped with mixed-valence impurities. *Technical Physics Letters*, 2015. Vol. 41, № 10. PP. 930–932.

[7] Karpova, S.S. Moshnikov, V.A. Mjakin, S.V. Kolovangina, E.S. Surface functional composition and sensor properties of ZnO, Fe_2O_3, and $ZnFe_2O_4$. *Semiconductors*, 2013. Vol. 47, № 3. PP. 392-395.

[8] Averin, I.A., Igoshina S.E., Moshnikov V.A., etc. Sensitive elements of vacuum sensors based on porous nanostructured SiO_2–SnO_2 sol–gel films. *Technical Physics*, 2015. Vol. 60, № 6. PP. 928–932.

[9] Averin, I.A. Igoshina, S.E. Karmanov, A.A. etc. Simulation of the sensor response of vacuummeters with sensitive elements based on multicomponent oxide nanomaterials with the fractal structure. *Technical Physics,* 2017. Vol. 62, № 5. PP. 799–806.

[10] Pronin, I.A. Averin, I.A. Yakushova, N.D. etc. Theoretical and experimental investigations of ethanol vapor sensitive properties of junctions composed from produced by sol-gel technology pure and Fe-modified nanostructured ZnO thin films. *Sensors and Actuators A: Physical*, 2014. Vol. 206. PP. 88-96.

[11] Pronin, I.A. Averin, I.A. Yakushova, N.D. etc. Investigation of milling processes of semiconductor zinc oxide nanostructured powders by X-ray phase analysis. *Journal of Physics: Conference Series,* 2017. Vol. 917. P. 032019.

[12] Dimitrov, D.Tz. Nikolaev, N.K. Papazova, K.I etc. Investigation of the electrical and ethanol-vapour sensing properties of the junctions based on ZnO nanostructured thin film doped with copper. *Applied Surface Science*, 2017. Vol. 392. PP. 95 – 108.

In: Applied Aspects of Nano-Physics …
Editors: Kirill Levine and Andrey Syrkov
ISBN: 978-1-53614-708-7

Chapter 41

NUMERICAL METHOD IN MODELING OF OBTAINING THIN FILM PROCESSES

V. A. Tupik, V. I. Margolin and Chu Trong Su*
Department of Microelectronics and Radio Engineering
of Saint Petersburg Electrotechnical University "LETI,"
Saint Petersburg, Russia

ABSTRACT

The high demand for improvement of the quality of obtained thin films leads to an indispensability of technological realization to use of numerical methods during optimization and synthesis of thin film structures. In the given work features of a numerical method in modeling processes of thin film formation are shown. Procedure of application of numerical methods for increase of accuracy and efficiency of computing processes was demonstrated.

Keywords: numerical method, computer modeling, thin film process technology

INTRODUCTION

Modeling is one of three compound components of any natural science. In nanoscience related to the Heisenberg uncertainty principle, the limitation of the resolution of the human eye, the randomness and self-organization of passing processes, the phase transition and the nonequilibrium state, the role of modeling plays an even

* Corresponding Author's Email: chusu171@mail.ru.

more significant role in the processes of synthesis, control and optimization of the process. The current state of the problem of modeling the process is to limit the linear scale from below (number of atoms) and from above (quantum effects). From classical physics to modern quantum physics, theoretical notions cover a space-time scale, but for a practical task, overlapping the entire range is not always good, because effectively implemented modeling should be well supported by experiments. Therefore, depending on the specific task presented, it is necessary to select an adequate numerical method. This or that numerical method is chosen usually based on a compromise solution between the presence of powerful computing facilities, accuracy, calculation time and the stability of the algorithm.

In connection with the compromise solution, a multiscale approach is currently widely used, namely that the objective function corresponds to macro- (roughness, thickness, application rate, pressure, temperature, geometrical parameters of the installation) and micro- (atomic structures, number and type of defects) to the parameters of the system and structure. In this paper, attention was paid to numerical methods, their application, and to the technique for realizing in the problem of modeling the processes of formation of thin films.

MODELS, NUMERICAL APPROACH AND DISCUSSION

To improve the quality of thin films, the configuration of the plant and the initial parameters were preliminarily calculated by the finite element method and by various methods for solving integral and differential equations (state-space representation) to obtain uniform thin film growth conditions (macro scale) [1, 2]. After this stage, a dynamic process for the particle system (atoms, molecules and clusters) is considered at the micro level for the purpose of calculating the preliminary characteristics of the process for a system of interconnected particles, after which the finite element methods and their modifications are returned to the calculation of the macro-mechanical characteristics of the obtained thin films [3], such as physical and heat-mechanical (temperature coefficient of expansion, porosity, roughness, mechanical modules), chemical bonds (agreement with the length of bonds, valence). For specific problems or materials where very high accuracy is required or there is no possibility of realizing many assumptions, a bottom-up approach based on quantum physics and on the first principles is used [4].

In connection with the above, an attempt to use a computer approach to the problems of synthesis and optimization of the technological (physical) process of thin film formation involves the use of the algorithm shown in Figure 1 [5], where the following

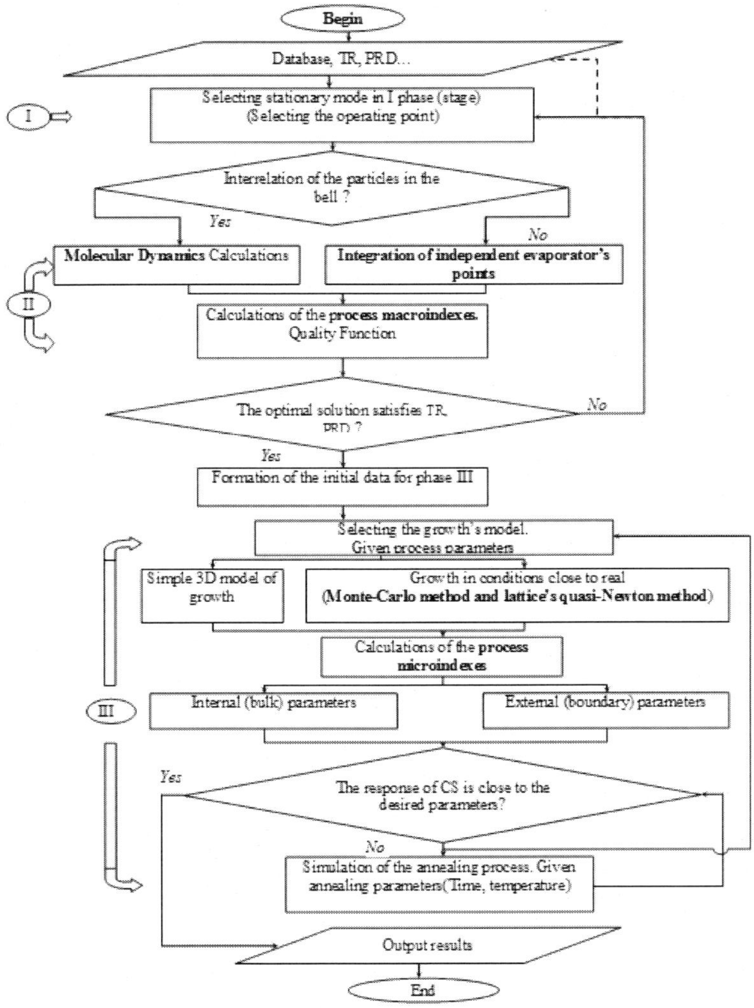

Figure 1. The calculated algorithm for synthesis and optimization of obtaining thin films technological (physical) processes.

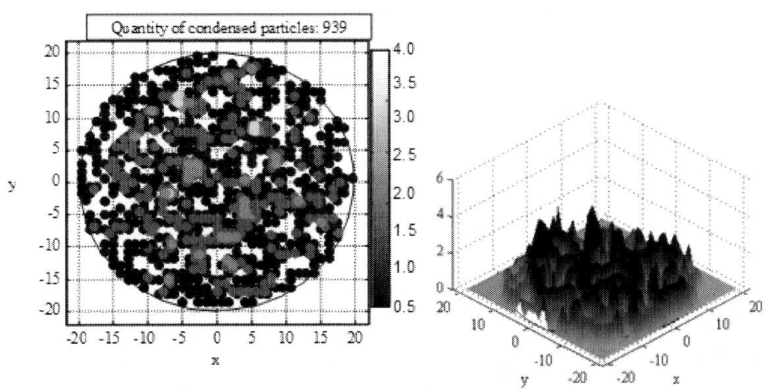

Figure 2. The top view of the reseived thin film from 939 particles in model of simple atoms (at the left) and the restored relief of a surface (on the right) [6].

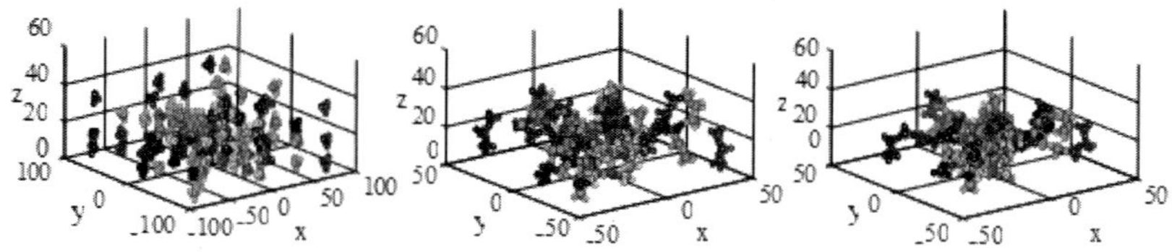

a) b) c)

Figure 3. Estimated computer simulation for 100 identical (complex) atoms in the initial (a), intermediate (b) and steady state (final structure) (c). Different colors of atoms and orbitals were chosen to improve the quality of visualization.

abbreviations are accepted: TR – technical requirements, PRD – product requirements document, CS – computer simulation, the first stage I – evaporation (or spattering, depending on the underlying technology), the second one II – mass transfer process from the target (or vapor source) to the substrate and in the end III stage is condensation and thin film growth process.

To simulate the behavior of particles in thin film growth processes, we use the Monte Carlo method (particle selection) and the quasi-Newton method (determination of the directions of particle motion), which uses, in describing the self-organizing actions of particles, their displacement only over the nodes of the coordinate (calculated) grid (which means quasi-method).

In Figure 2 shows the result of computer simulation for simple atoms, which are described only by isotropic potential interconnection. In this case Lenard-Jones pair potential was chosen, more detail is shown in [6, 7].

The above methods can't be used for atoms that are bound by anisotropic bonds. For instance semiconducting material, it is inseparable part of micro-nano electronic. In this connection, in [8, 9] an attempt is made to improve the empirical approach on the basis of the proposed new models of one atom and interatomic interaction with sp3 type of hybridization.

In Figure 3 shows the result of a trial computer simulation, this model, allowing to take into account the lengths, angles and anisotropy of the bonds between two particles.

As it's shown the necessary to take into account, depend on the demand of task to get some compromise solution with permissible accuracy, more detail is shown in [9].

Conclusion

Realization of new possibilities and features of films and coatings is associated with the demand for new computer techniques and technologies, it is necessary to monitor the capabilities and limitations of available computing facilities, numerical methods and calculation algorithms. Understanding of the problem and the correct choice of the mathematical method or a combination thereof, among many basic numerical methods and their modification leads to an adequate description of the ongoing processes, which makes it possible to carry out computer experiments and obtain valuable predictions. In this paper, we showed the results of using the original approaches of numerical methods on a personal computer to synthesize and optimize the process of formation of thin films.

ACKNOWLEDGMENTS

The authors would like to acknowledge the financial support of this work by the Ministry of the Education and Science of the Russian Federation (basic state assignment No. 8.7130.2017).

REFERENCES

[1] Song Ruiliang, Liu Wei, Zhang Yi, Cai Yongan, Sun Yun. Finite element simulation and experimental research on ZnO:Al by magnetron sputtering. *Thin Solid Films*, 2011. – Vol 502. – Issue 2. P- 887 - 890.

[2] Tupik, V. A., Chu Trong Su, I. Steblevska, "The simulation and the optimization of the quality function of the process of formation of functional coatings and films", *Proceeding of the Russian Universities: Radioelectronics*, 2015. – N. 5.15 - 19 (in Russian).

[3] Amir R. Khoei. *Extended finite element method: theory and applications*. John Wiley & Sons, Ltd., 2015. – 600 p. – ISBN: 1118457684.

[4] Javier Junquera and Philippe Ghosez. Critical Thickness for Ferroelectricity in Perovskite Ultrathin Films. *Nature,* 422(6931):506-9 May 2003 DOI: 10.1038/nature01501.

[5] Tupik, V. A., Margolin, V. I., Chu Trong Su. Computer Experiment in Obtaining Thin Films Process Technology: *Tool for Modeling and Training. Science, Education, Innovation,* Saint-Petersburg, Russia, 2016. pp. 43-45. Doi: 10.1109/IVForum.2016.7835849.

[6] V. A. Tupik, V. I. Margolin and Chu Trong Su. Consideration of the condensation processes of thin films in the crystal substrate's potential field. *J. Phys.: Conf. Ser.,* 729 012025.

[7] Chu Trong Su, "Computer Simulation of the Thin Film's Growth Process during Thermal Vacuum Evaporation", *Proceeding of the Russian Universities: Radioelectronics*, 2016. – N. 6.22-31. (in Russian).

[8] V. A. Tupik et al. *Smart Nanocomposites*, Volume 8, Number 1, p. 1-6, ISSN: 1949 - 4823.

[9] V. A. Tupik et al. 2017 *J. Phys.: Conf. Ser.,* 872 012052.

In: Applied Aspects of Nano-Physics ...
Editors: Kirill Levine and Andrey Syrkov
ISBN: 978-1-53614-708-7

Chapter 42

INVESTIGATING ORDERING DEGREE OF NON-CRYSTALLINE FILMS IN HETEROJUNCTION SILICON SOLAR CELLS BY RAMAN SPECTROSCOPY

Aleksei D. Maslov**, Ekaterina V. Bezuglaya and Nikolay V. Vishnyakov
Department of Micro- and Nanoelectronics,
Ryazan State Radio Engineering University,
Ryazan, Russia

ABSTRACT

The heterojunction silicon solar cells based on a-Si/c-Si was investigated by Raman spectrocopy. We analyzed Raman spectra to determine structural characteristics of amorophous film that allows to make a conclusion about degree of disordering of amoprhpous silicon layer on the basis of estimating angle of coupling deviation. It is shown that degree of disordering strongly depends on technological conditions of manufacturing amorphous layer. Correlations between such technological parameters as substrate temperature, deposition time, plasma treatment of surface and structural properties of a-Si are found.

Keywords: heterojunction silicon solar cells, Raman spectroscopy, structural characteristics, technological conditions, amorphous layer

* Corresponding Author's Email: Maslov.a.d@mail.ru.

INTRODUCTION

Heterojunction solar cells based on a-Si/c-Si junction are promising structures for effective cheap terrestrial use. However disorder amorphous layers might restrict efficiency of such structures due to the recombination by dangling bonds, parasitic absorbance and resistivity of these layers. Therefore it is necessary to control such parameters of a-Si layers as thickness, doping concentration, density of recombination states etc. Since density of states strongly depends on structural features of amorphous web it is necessary to investigate and control structural characteristics.

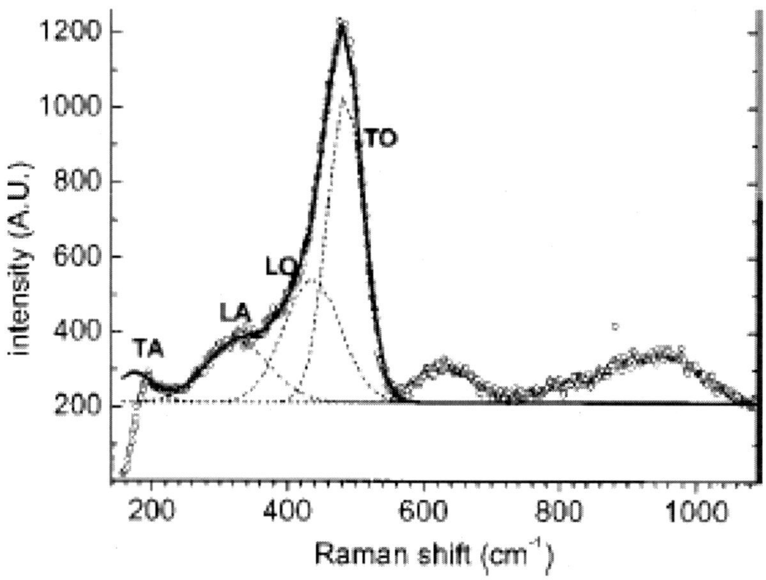

Figure 1. Typical spectrum of amorphous silicon [2].

These parameters depend on technological conditions of amorphous layer manufacturing. Thus the aim of this paper is determination of correlations between technological conditions and structural characteristics of a-Si.

For investigation of structural characteristics we used widespread Raman spectroscopy method that is often used to study structural information [1]. Raman spectrum of amorphous silicon is a complicated spectrum that consist of four peaks corresponding TA - transverse acoustical, TO - transverse optical, LO - longitudinal acoustic and LO - longitudinal optical vibrations of non-crystalline web. Typical spectrum of amorphous silicon is shown on Figure 1 [2]. Compared to other methods of investigating structural properties, Raman spectroscopy is a more sensitive method to small fluctuations in structure within the first coordination sphere [3]. For example, measuring width of TO-peak allows to calculate root-mean-square deviation of the coupling angle from the tetrahedral position of the silicon crystal using Beeman's linear empirical relation [3]:

$$\Gamma = 15+6\Delta\theta \tag{1}$$

where Γ- width of the peak, cm^{-1}; $\Delta\theta$ - root-mean-square deviation, degrees.

EXPERIMENTAL RESULTS

We investigated four testing a-Si:H/ c-Si solar cells manufactured with different technological conditions described in Table 1.

Raman spectra are obtained using automated AFM-Raman, SNOM and TERS system Ntegra Spectra (NT-MDT, Russia). We used a 532nm solid state laser with power 0.4mW.

The obtained Raman spectra are shown in Figure 2 and 3.

In addition to Beeman's empirical relation (1), the degree of ordering might be determined by estimation of TO-peak position and the ratio of TA/TO peak intensities [3].

Table 1. Technological conditions of amorphous silicon

№	Substrate temperature,°C	Deposition time of a-Si:H, hours	Thickness. um	Notes
1	220	5	3	
2	170	5	3	
3	230	2,5	1,5	
4	230	2,5	1,5	Plasma treatment of surface

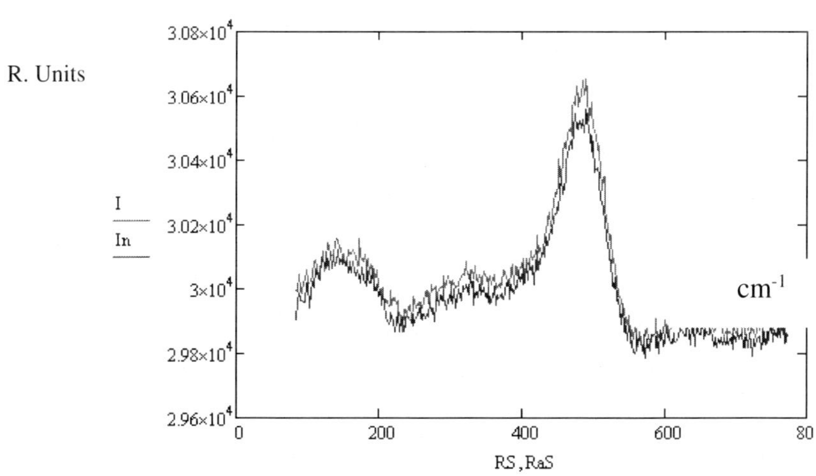

Figure 2. Raman spectra of samples №1 (I) and №2 (In).

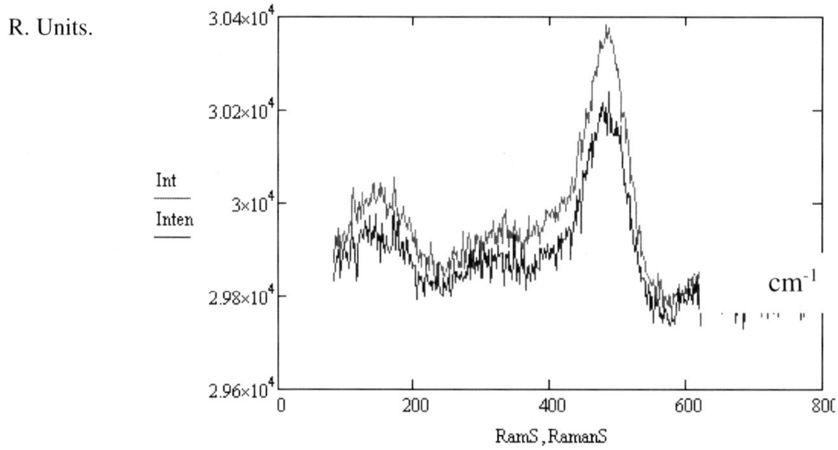

Figure 3. Raman spectra of samples №3 (Int) and №4 (Inten).

Results of calculating the ordering degree of amorphous layer are shown in Table 2.

Table 2. Results of calculating structural parameters

Parameter Sample	Angle of coupling deviation, degrees	TO-peak position, cm^{-1}	Ratio of TA/TO peak intensities
1	8,6	486	0,983
2	8,8	486	0,986
3	10,6	486	0,988
4	10,8	474	0,991

Disordering of amorphous web should influence on increasing of deviation of the coupling angle from the tetrahedral position appropriate to crystalline silicon, the intensity ratio of the TA/TO peaks, and the shift of the TO peak position. Thus changes in structural characteristics of the samples are correct [4]. The samples №1 and №2, manufactured with different substrate temperature have different angle of coupling deviation and TO/TA peak ratio: sample №1 has lower values because this film is less disordered due to the lower substrate temperature (Table 1).

Increasing of angle of coupling deviation and TO/TA peak ratio of samples №3,4, obtained with higher substrate temperature than samples №1,2, is related with influence of deposition time on formation of amorphous web.

In the case of comparison of samples №3 и №4, manufactured with substrate temperature 230°C, sample №4 with surface bombarded by Ar^+, according to the Table 2 is more disorder film due to the increase of defects states. It is confirmed by shift of TO-peak.

CONCLUSION

We found correlations between technological conditions of amorphous silicon layer and structural characteristics using Raman spectroscopy method. Substrate temperature, deposition time and plasma treatment of surface are considered. The degree of disordering of amorphous web is estimated by Beeman's empirical relation, TO-peak position and the ratio of TA/TO peak.

The experiments are made using equipment of Regional center of probe microscopy of Ryazan state radio engineering university.

REFERENCES

[1] Ferraro, John R., Kazuo Nakamoto and Chris W. Brown. *Introductory Raman spectroscopy*. Elsevier, 2003. p. 434.

[2] C. Smit, R. A. C. M. M. van Swaaij, H. Donker, A. M. H. N. Petit, W. M. M. Kessels, M. C. M. van de Sanden. Determining the material structure of microcrystalline silicon from Raman spectra. *J. Appl. Phys.,* 2003. Volume. 94. pp. 3582 - 3588.

[3] R. L. C. Vink, G. T. Barkema, W. F. van der Weg.. *Relation between Raman spectra and Structure of Amorphous Silicon,* 2000. Vol. 19. pp. 156 - 162.

[4] Brodsky, Marc H. *Amorphous semiconductors*. Springer-Verlag, 1979. p. 337.

In: Applied Aspects of Nano-Physics …
Editors: Kirill Levine and Andrey Syrkov

ISBN: 978-1-53614-708-7

Chapter 43

COMPOSITE NANOSTRUCTURED MATERIALS FOR PLASMA ENERGETIC SYSTEMS

***R. S. Smerdov*[1,*], *A. S. Mustafaev*[1], *Yu. M. Spivak*[2] and *Vyacheslav A. Moshnikov*[2]**

[1]Saint Petersburg Mining University, Saint Petersburg, Russia
[2]Saint Petersburg Electrotechnical University "LETI", Saint Petersburg, Russia

ABSTRACT

The possible applications of composite nanostructured materials for low work function PETE (photon-enhanced thermionic emission for solar concentrator systems) and thermionic plasma energy systems synthesis is discussed in this chapter. A wide range of semiconductor, graphene and fullerenol based materials is suggested for further development and study due to unique physical effects existing in these structures. The nature of such phenomena is discussed and interpreted as due to the influence of functionalization and synthesis parameters on the structure properties of composite materials.

Keywords: plasma energetic systems, porous silicon, nanostructured graphene layers, fullerenol, PETE, solar concentrators

[*] Corresponding Author's Email: rostofan@gmail.com.

INTRODUCTION

Thermionic energy converters (TECs) are promising heat engines allowing the straightforward conversion of thermal energy into electricity. A typical thermionic converter (TEC) comprises an emitter (cathode) operating at an elevated temperature and a collector (anode) kept at a lower temperature separated from the cathode with a vacuum gap. Only a fraction of the electrons possess enough thermal energy to surmount the TEC emitter material's work function and escape into vacuum, thus producing current between the two electrodes. The Richardson law governs the thermionic emission current density according to: $J = A_c{*}T_c c^2 e^{-\varphi c/kTc}$, where φ_c is the electron work function from the cathode surface, T_c is the cathode temperature, A_c is the Richardson's constant for a particular material.

Thermionic converters were initially developed and manufactured in the mid-1950s, their conversion efficiencies reaching approximately 10–15% [1]. Both the US NASA and the Soviet space programs endorsed further development of TECs for autonomous satellite energy sources and other applications requiring relatively high-power independent energy generators. It is worth noting that TEC efficient operation requires the creation of materials capable of withstanding high temperatures and significant current densities thus providing a solid foundation in particular fields of material science research.

The goal of this study is to identify and compare the possible methods of energy conversion while suggesting promising solutions and developing novel plasma energy systems' electrode materials that will further facilitate energy conversion efficiency growth.

PETE SOLAR CONCENTRATORS

The solar energy conversion to electricity can be realized by utilizing a number of techniques including the "quantum" approach currently realized in photovoltaic cells allowing the appliance of solar energy to excite electrons, and the "thermal" approach, for which concentrated sunlight is used as a source of thermal energy generation using various types of heat engines. In practice, the devices based on the two approaches combination rapidly lose their effectiveness with both excessive elevation of operating temperatures resulting in the rapid degradation of the photocells and with lowering of operating temperatures reducing the efficiency of thermal engines. Utilizing the systems based on the phenomenon of photon-enhanced thermionic emission (PETE) allows to overcome these drawbacks by simultaneously realizing the photovoltaic and thermionic effects that allow the DI of both quantum and thermal mechanisms in one physical process [2]. The study of electron emission from composite nanostructured materials into vacuum presents both great scientific and practical value. The scientific interest resides in elucidating the mechanisms of electron transport to the surface and subsequent emission into vacuum, identifying particular potential barriers at the interfaces, such as band bending and electron affinity, resulting in further understanding of their influence on the quantum yield. Practical significance comprises the improvement of electrode parameters leading to an increase in energy conversion efficiency. It is well known that the deposition of monoatomic layers of alkali and alkaline earth metals on the pure surface of an electrode leads to a significant decrease in electron affinity due to the formation of surface microdipoles. As a consequence, the potential barrier is reduced and the quantum yield of emission is increased. With the simultaneous adsorption of cesium and oxygen on the electrode surface, it is possible to achieve energy states characterized with a negative effective electron affinity (NEA) in which the vacuum level (E_{vac}) lies lower than the bottom of the conduction band (E_c) in the volume of the material. Due to the state characterized with NEA, the spectral threshold of photoemission is reduced to approximately ~ 1.4 eV (for example the band gap (E_g) of GaAs is 1.4 eV), which corresponds to the near infrared region, and the quantum yield reaches approximately ~ 50% [2]. In this connection, NEA cathodes based on p-GaAs (Cs, O) are widely used in photoelectron multipliers and electro-optical converters and solar concentrator systems [2].

Emission from semiconductors is usually described using a "three-stage" model, including photoexcitation of electrons from the valence band to the conduction band, transport of electrons to the emitting surface, and the emission process – transition of electrons through the semiconductor-vacuum boundary. While the first two stages – photoexcitation and transport in the volume of the semiconductor are relatively well understood, the third stage – the actual emission process itself, has not been studied enough, in spite of many years of NEA photocathodes studies. The microscopic mechanisms determining the probability of electron exit into vacuum remain undefined, which, in turn, limits the quantum yield of NEA photocathodes in comparison with the theoretical limit (100%). Moreover, it has been established that the probability of electron emission into vacuum in a single collision with an NEA surface is small (much less than one), that is, significantly less than the probability of passing the NEA boundary approximated by a negative potential step using the effective mass approximation [3].

The reasons underlying such miniscule values of the emission probability under condition of a single collision of an electron with a NEA surface have not yet been established in microscopic scale. One of the possible reasons is the presence of a narrow potential barrier on the GaAs – CsO – vacuum interface through which the electron has to tunnel in order to escape into the vacuum. Other possible reasons include the scattering of the momentum and electron energy in the (Cs, O) layer, as well as the effective mass approximation being inadequate to calculate the probability of electrons passing through interfaces. However, regardless of these reasons, the question arises as to how the probability of an electron escape into vacuum for a NEA surface turns out to be significant (up to ~ 0.5) if, at each collision, an electron emits into the vacuum with such a low probability. An explanation of this paradox was proposed in [4]. According to [4], in the region of surface zone bending, electrons are captured in a size-quantization effect associated subband whose bottom is located above the vacuum level. The resulting probability of an electron leaving the surface into the vacuum increases due to its multiple collisions with the NEA surface during electron's lifetime in the subband. As a result, the quantum yield from NEA-photoemitters can reach a value slightly higher than 50%. However, even such large values of the quantum yield are much lower than the theoretical limit of 100% and, therefore, a significant fraction of the photoelectrons recombine on the surface.

Recently, semiconductor surfaces with a small positive electron affinity (PEA) and relatively low electron work function have also attracted lots of attention due to the possibility of increasing the solar energy conversion efficiency through the use of "photon-enhanced thermionic emission" (PETE) electrons generated by the incident light characterized with the energies below the vacuum level and thermalized due to PETE-effect [2].

The prototype proposed in [2] is based on the effect of thermionic emission of photoexcited electrons from a semiconductor cathode at high temperatures. The convertor operated at elevated temperatures (higher than 200°C), which allowed the possibility of thermal energy utilization in order to realize the subsequent thermionic emission process, increasing the theoretical efficiency of the combined conversion to the values of approximately 50% [2].

The possibility of PETE based systems synthesis with semiconductor (GaN) electrodes was demonstrated in [2], however these prototypes' efficiency is significantly hindered by a crucial drawback: the number of incident photons characterized with energies exceeding the band gap of GaN (Eg = 3.3 eV) is less than 1% of their total quantity. That is why, the studies of alternative structures, including but not limited to porous silicon (PS) and PS-based nanocomposites for subsequent electrode synthesis are promising, since the band gap of such materials can be varied in a wide range from 1 to 3 eV due to the existence of quantum confinement effect [5] and significant opportunities for surface functionalization [6,7,8]. The phenomenon of photon-enhanced thermionic emission was also studied experimentally for Cs/p^+-GaAs system [9]. According to [2] such system is suited for the creation of PETE solar energy converters, because the value of GaAs band gap lies in close proximity to the optimum and by changing the properties of the cesium coating, the optimal value of the electron work function from the surface can be achieved. In these circumstances in addition to the emission of hot electrons generated by light characterized with the energies above E_{vac}, another mechanism of electron emission exists. Electrons generated by interband light and possessing energies below the vacuum level can be scattered by lattice vibrations (phonons) leading to the activation of a thermalization process. In thermodynamic equilibrium, thermalized electrons can be described with the Maxwell distribution, and hence, the emission of electrons is possible directly from the tail of the distribution function.

In order to maximize the emission effectiveness the use of materials with controllable work function is required. The p-type conductivity semiconductors with low electron affinity in which the Fermi level lies near the top of the valence band are suitable for further development as the PETE anodes. In this case the relation between effective electron affinity (χ^*) and the electron work function (φ) is $\chi^* + E_g \approx \varphi$, where E_g is the band gap of a particular material. The effective electron affinity can be expressed as: $\chi^* = \chi - \varphi_s$, where φ_s is the band bending due to the existence of unoccupied states on the surface and χ is the true electron affinity of the material. Thus, in order to reduce the effective electron affinity, it is necessary to reduce the true electron affinity and increase the bending (φ_s) of the bands. The true electron affinity can be reduced by creating the monoatomic layers of alkali and alkaline earth metals or by utilization of various molecules characterized with a large dipole moment on the semiconductor surface.

THERMIONIC ENERGY CONVERTERS

In consideration of energy applications for both thermionic energy converters (TECs) and PETE solar concentrators the optimal value of φ_c reflects an intrinsically complicated balance between maximizing electron emission current and operating voltage, while the anode work function φ_a should be kept as low as possible in order to maximize theoretical output voltage. In practice, this fact indicates that the anode temperature should be kept low enough to minimize reverse thermionic emission current.

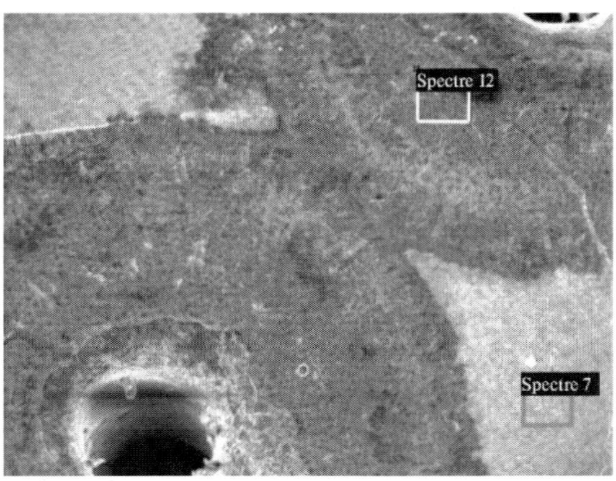

Figure 1. SEM image of the collector's surface (secondary electron detection mode) at 20 keV beam energy [10].

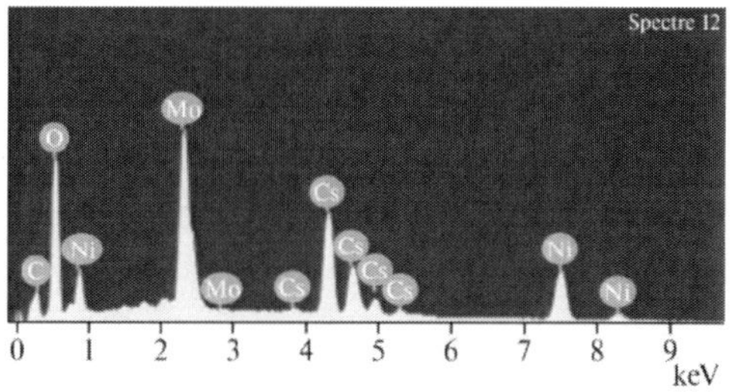

Figure 2. X-ray spectrum of the "Spectre 12" region in Figure 1. The ratio of atomic concentrations of the elements C : O : Ni : Mo : Cs – 27% : 53% : 12% : 2% : 10% [10].

It is well known that the synthesis of plasma energy systems based on thermionic emission requires the creation of materials with a low work function of electrons from the surface (φa). The electromotive force produced by thermionic energy converters is mainly determined by the electron work function (φc) of the cathode (emitter) material, while the output voltage loss is related to the electron work function (φa) of the anode (collector) material. Since a significant increase in the emitter electron work function by selecting a suitable material is ineffective due to the reduction in the generated electric current density (in accordance with Richardson's law) at the technologically specified emitter temperature (at which the evaporation rate of the cathode material remains acceptable), it is necessary to reduce the anode electron work function φa in order to improve the TEC energy efficiency.

The problem of reducing φa is usually solved by utilizing alkali and alkaline-earth metal coatings, in particular, cesium (Cs) [10]. For example, anodes based on tungsten coated with cesium are traditionally used because of high thermal stability and a relatively low work function (1.7 eV). The use of a nickel anode coated with graphene layers intercalated with cesium atoms made it possible to obtain an unprecedented decrease in the work function of electrons from the surface of the material (φ_a <1 eV) and, consequently, a threefold increase in the energy conversion efficiency (up to 25%) [10].

Figure 2 shows the X-ray spectrum of the region designated as "Spectre 12" on Figure 1. The appearance of a considerable amount of Cs atoms in the area within the darker concentric rings could be noted whereas the area outside those rings does not contain any Cs as one could clearly see from Figure 3 (X-ray spectrum of the "Spectre 7" region). This can be attributed to the fact that Cs atoms, being only physically adsorbed on the surface of the original graphite coating, were pumped out after the cooling process of Cs reservoir. In contrast, the region of Figure 2 represented by areas within the concentric rings in Figure 1 is characterized with Cs atoms intercalated in the space between the graphene layers comprising the structured graphite grains.

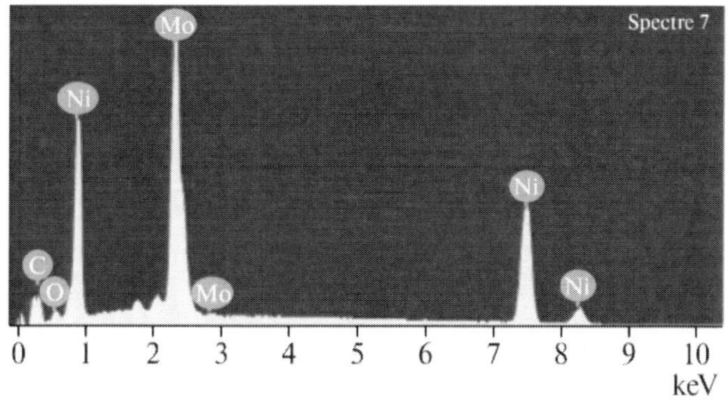

Figure 3. X-ray spectrum of the "Spectre 7" region in Figure 1. Ratio of atomic concentrations of the elements C: O : Ni : Mo : Cs – 47% : 5% : 27% : 21% : 0 [10].

The steep increase in the number of O atoms, (represented on the X-ray spectrum in Figure 2) is conditioned by the fact that the oxygen atoms practically "accompanied" Cs atoms while simultaneously oxidizing them thus creating a chemical bond. The bond between the O atom and the two Cs atoms one of which intercalated between graphene layers and the other one arrived from the Cs vapor resulted in the overall dipole moment increase thus diminishing the electron work function value [10].

Precipitation of alkali and alkaline earth metals is not the only method of modifying the material's electron work function. In particular, the use of high electron affinity PS and PS-based nanocomposites functionalized with silver nanoparticles (Ag NPs) and C_{60}-Ag fullerenols as low work function anodes for energy systems appears to be very promising due to the fact that the chemical bond of the C_{60} molecule to the Ag atom leads to an increase in the dipole moment (caused by electron density displacement towards the C_{60} molecule) of the anode surface as a whole. In this case, a partially filled layer of dipoles with its positive charge heading away from the surface to the outer half-space emerges. This particular orientation of surface dipoles causes a decrease in the potential barrier thus leading to a significant decrease in electron work function allowing to consider PS-based composite structures as a promising material for plasma energy electrode synthesis.

CONCLUSION

PS and PS-based nanocomposites due to their intrinsic properties show a lot of potential as perspective materials for PETE and TEC energy convertors electrode synthesis. Although, the existing structures show acceptable performance, allowing to achieve relatively high conversion efficiencies, further studies are required in order to improve a wide range of their characteristics including, but not limited to electron work function variability, durability and thermal stability as well as to realize a facile, reliable, controllable and scalable synthesis technique.

Current developments it the field of low work function materials allowed the synthesis of a novel composite nanostructured material: Cs-intercalated graphene coated Ni based electrode (anode). The creation of this material allowed to increase the energy conversion efficiency dramatically (up to 25%), providing a solid foundation for further research and improvement of thermionic energy converters, although the application of conventional methods to study surface physics under conditions of a significant concentration of cesium vapor and high temperatures is rather intricate.

REFERENCES

[1] Hatsopoulos G.N., E.P. Gyftopoulos. *Thermionic Energy Conversion* Vol. I. 1973. MIT Press. 265P.

[2] Schwede J.W., I. Bargatin, D.C. Riley, B.E. Hardin, S.J. Rosenthal. *Nature Materials.* 2010. Vol. 9. P.762-767.

[3] Karkare S., D. Dimitrov, W. Schaff, L. Cultrera, A. Bartnik, X. Liu, E. Sawyer, T. Esposito, and I. Bazarov, *J. Appl. Phys.* 2013. Vol. 117. 104904.

[4] Korotkikh V.L., A.L. Musatov, V.D. Shadrin. 1978. *JETP Lett.* Vol.27 P.617

[5] Nolan M., S. O'Callaghan, G. Fagas, J.C. Greer. *Nano Lett.* 2007. Vol. 7. №1. P. 34-38.

[6] Smerdov R.S., V.V. Loboda, Yu.M. Spivak, V.A. Moshnikov. *SPbSPU Journal. Computer Science. Telecommunication and Control Systems.* 2016. № 3 (247). P.13-22.

[7] Spivak Yu.M., S.V. Mjakin, V.A. Moshnikov et al. *J. of Nanomaterials.* 2016. Vol. 8. 8P.

[8] Mustafaev, R., Smerdov, B. Klimenkov. *Bulletin of the American Physical Society.* 2017. Vol.62. №12.

[9] Madey T.E., J.T. Jr Yates. *J. Vac. Sci. Technol.* 1971. Vol. 8. P.525-555.

[10] Mustafaev A.S., V.A. Polishchuk, A.B. Tsyganov, V.I. Yarygin, P.A. Petrov. *Russian Journal of Physical Chemistry* B. 2017. Vol. 11. №1. P.118-120.

In: Applied Aspects of Nano-Physics ...
Editors: Kirill Levine and Andrey Syrkov
ISBN: 978-1-53614-708-7

Chapter 44

EFFECTS OF GRAFITE INTERCALACATION WITH CESIUM IN A THERMIONIC CONVERTER

A. S. Mustafaev*

Saint Petersburg Mining University, Saint Petersburg, Russia

ABSTRACT

The thermionic energy converter (TEC) with inter-electrode low-temperature plasma and cesium vapor dynamic flow in the inter-electrode gap (IEG) demonstrated an increase in efficiency of up to 20%. This was mainly reached due to the decrease in the effective electron work function to an anomalously low value of the order of 1 eV from a perforated nickel collector covered with nanosized graphite flakes under the converter working conditions. SEM X-ray microanalysis of the collector surface layers was performed, and the model of the effects proposed. In a such kind of collector of TEC with a nanosized graphite coating, multilayered intercalation of Cs atoms in the graphite/graphene surface layers was found during work in cesium vapors at a pressure of ~ 1–10 Torr. This probably explains the anomalously low electron work function and the high efficiency of TEC.

Keywords: thermionic energy converter, electron work function, nickel collector, nanosized graphite, elemental microanalysis, graphite intercalation by cesium atoms

* Corresponding Author's Email: ale*xmustafaev@yandex.ru.

INTRODUCTION

Studies on thermionic energy converter (TEC) performed at the Institute for Physics and Power Engineering and aimed at seeking methods for drastically raising the efficiency (several-fold) and operating life (by approximately an order of magnitude) relative to the previously achieved level (~10% and ~1 year, respectively) in TOPAZ and Yenisei space nuclear power units [1] showed that it is possible to achieve high efficiency of converters with electrodes of advanced materials [2]. An efficiency of ~20% was obtained for the first time using an experimental TEC under laboratory conditions in an electric energy generation mode with dynamic feeding of cesium vapor in the inter-electrode gap (IEG) at an emitter temperature of T_E ~ 1600 K and collector temperature of T_C ~ 700 K. If reproduced in electrogenerating systems with an emission surface of the order of several square meters, the unique combination of relatively low temperature T_E and high efficiency of this thermionic system opens up prospects for introducing these systems in ground autonomous small-scale nuclear power engineering with direct conversion and co-generation in thermal power plants of conventional composition. As is well known, the electromotive force developed by TEC is mainly determined by the electron work function (EWF) of the emitter material, and the output voltage losses depend on the EWF of the collector material and plasma voltage losses in the IEG. The efficiency of TEC can be increased mainly by decreasing the collector EFW. Earlier [3, 4], high efficiency of laboratory TEC was demonstrated experimentally by continuously circulating cesium vapors through a nickel collector with many holes with a diameter of 0.1 mm. The effect was observed if the collector surface was coated with graphite using an aquadag suspension, and not observed on a metal surface.

RESULTS OF EXPERIMENT

The experiments performed in [3, 4] were repeated independently on an improved laboratory TEC with a perforated collector with a coating of nanosized graphite flakes kindly supplied by one of the authors of [3, 4]. The study was performed with planar electrodes and with a variable IEG (0.2–3 mm) using a vacuum path for feeding cesium, which allowed either equilibrium or dynamic feeding of cesium vapors in one experiment [2]. As TEC electrodes, we used an emitter of vacuum-melted polycrystalline Mo with a diameter of 14 mm and a thickness of 11 mm with a Pt (~3 μm) coating and a collector (with a diameter of 8 mm) of laser-perforated nickel foil with a thickness of 0.2 mm (121 holes with a diameter of 0.1 mm on a square of 4 × 4 mm). The emitter was heated by electron bombardment; $T_E = 1350$ K, $T_C = 750$ K, $T_{Cs} = 570$ K.

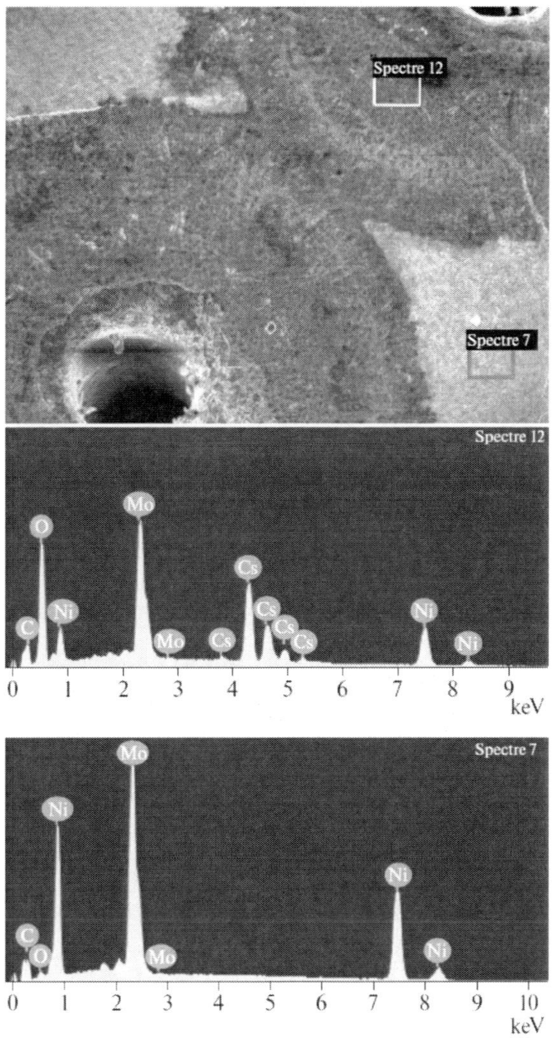

Figure 1. SEM image and X-ray fluorescence spectra of the surface of a perforated nickel collector coated with nanosized graphite after prolonged exposure to Cs vapors.

The current-voltage characteristics (CVCs) of TEC were measured by the pulse method using electric current scanning from a static working point on the CVC corresponding to the diffusion mode of operation. The CVCs were measured and optimized both with equilibrium feeding of cesium vapors and after repeated activation of TEC electrodes using the conditions suggested in [4, 5]. The measurement time was at least 1000 h.

The experimental CVCs may be explained if the electron work function of the collector does not exceed 1 eV. As is known, all metals or alloys have an EWF of a few electron-volts; at room temperature, the only composite with a work function as low as this is the silver-oxygen-cesium photocathode [6], whose photoemission threshold is around 1200 nm (~1 eV). Therefore, it was interesting to determine the state of the surface of a perforated collector of Ni coated with nanosized graphite flakes after prolonged work in a cesium atmosphere of the described TEC. For this purpose, after the experiments were completed, the cesium vapor was pumped out, the collector was cooled and extracted from TEC, and its surface was analyzed on a scanning electron microscope (SEM Merlin, Zeiss) with an Oxford Instruments X-Max energy dispersive X-ray detector.

Table 1. Contents (%) of elements in the surface layer with a depth of 0.9 μm of a Ni perforated collector with a nanosized graphite coating

Element	Around holes in diffusion rings	Outside the diffusion rings
C	22	37
O	46	4
Ni	14	41
Mo	9	18
Cs	9	0

The panoramic SEM images of the collector surface (beam energy 20 keV, secondary electrons mode) show concentric rings around each hole in the collector (Figure 1). They evidently suggest that diffusion occurs on the collector surface from the boundary of the hole. The experiments reported in [7] showed that the effect of lower work function does not appear in the absence of holes in the nickel collector even in the dynamic feeding mode of cesium vapors in IEG. Then the distribution of elements between the neighboring holes was measured in different regions of the collector based on the X-ray spectra excited with fast electrons from the microscope beam and recorded with an Oxford Instruments detector. Spectrum 12 from the region of the concentric diffusion ring in the figure showed Ni, Mo, C, Cs, and O lines. During the operation of TEC, molybdenum evaporated from the emitter and fell on the collector. In spectrum 7, there were no Cs and O X-ray lines outside the limits of the concentric diffusion ring in the figure. Monte Carlo simulation of fast electron trajectories showed that the depth of collector surface microanalysis was of the order of 0.9 μm in all cases.

The average atomic concentrations in the two zones (in the concentric rings around the holes and outside the rings) are listed in the table. Nearly equal partial concentrations of Cs and O atoms were observed at different points of the diffusion rings around the collector holes, but not observed outside the holes. This may be explained by diffusion (intercalation) of these atoms from the hole boundaries into surrounding multistoreyed (with a height of ~1 μm) layers of nanosized graphite–graphene. The observed Ni concentration inside the rings is lower than outside because Cs and O intercalated into multilayered graphite–graphene screened the lower-lying Ni atoms of the substrate from fast electrons.

Conclusion

In a perforated nickel collector of TEC with a nanosized graphite coating, multilayered intercalation of Cs atoms in the graphite/graphene surface layers was found during work in cesium vapors at a pressure of ~1–10 Torr. This probably explains the anomalously low electron work function and the high efficiency of TEC.

References

[1] Gryaznov G. M. and V. Ya. Pupko, *Priroda* (Moscow, Russ. Fed.), No. 10, 29 (1991).

[2] Yarygin V. I., *J. Clust. Sci.* 23, 77 (2012).

[3] Holmlid L., in *Proceedings of the Thermionic Energy Conversion Specialist Conference*, Göteborg, Sweden, 1993, p. 47.

[4] Svensson R., L. Holmlid, and E. Kennel, in *Proceedings of the Thermionic Energy Conversion Specialist Conference,* Göteborg, Sweden, 1993, p. 93.

[5] Svensson R. and L. Holmlid, in *Proceedings of the 32nd Intersociety Energy Conversion Engineering Conference*, New York, 1997, p. 1071.

[6] Sommer H., *J. Appl. Phys.* 51, 1254 (1980).

[7] Yarygin V. I., V. N. Sidel'nikov, I. I. Kasikov, V. S. Mironov, and S. M. Tulin, *JETP Lett.* 77, 280 (2003).

In: Applied Aspects of Nano-Physics …
Editors: Kirill Levine and Andrey Syrkov
ISBN: 978-1-53614-708-7

Chapter 45

CONTROL OF NONLINEAR PLASMA OSCILLATIONS IN A SHORT DC DISCHARGE WITH NONLOCAL IONIZATION

A. S. Mustafaev[1,*], A. Y. Grabovskiy[1*], V. I. Demidov[2], I. Kaganovich[3] and M. E. Koepke[2]
[1]Saint Petersburg Mining University, Russia
[2]West Virginia University, Morgantown, West Virginia, US
[3]Princeton Plasma Physics Laboratory, Princeton, New Jersey, US

ABSTRACT

A dc discharge with a hot cathode is subject to current and voltage plasma oscillations, which have deleterious effects on its operation. In this paper, we demonstrate how it is possible to achieve reliable suppression of the discharge oscillations by installing an auxiliary electrode, placed outside of anode. By collecting a modest current through a small opening in anode, we show that the discharge becomes stable, in a certain pressure range. This method of avoiding current oscillations can be used for wide range of applications.

Keywords: plasma chemistry, dc discharge, electron distribution function, plasma instabilities negative resistance, heated cathode

Gas-discharge plasma devices are widely used in plasma applications [1]. A better understanding of the physics behind these devices has been developed over the last few

* Corresponding Author Corresponding Author's Email: schwer@list.ru.

decades and has allowed for the creation of discharge plasmas with more controllable parameters, including modification of the charged particle densities, temperatures (average energies), and electron energy distribution functions (EEDF). However, improving the performance of these new plasma tools is still a significant challenge for plasma engineering. Exploitation of nonlocal plasma properties allows additional dimensions and flexibility in adjusting plasma parameters. A remarkable property of such plasmas is that changing conditions in one place may lead to unexpected changes far away in another part of the plasma. Additionally, plasmas with nonlocal EEDF [2, 3] allow independent and effective managing of electrons belonging to different energy ranges [4, 5]. This, in turn, allows modification of the plasma properties in desirable ways, because different energetic groups of electrons are responsible for different processes, and their density modifications yield control over corresponding plasma processes.

One example of such a device with nonlocal plasma properties is a short dc discharge (several millimeters in length at the pressure of a few Torr, 10–100 μm for atmospheric pressure) [6]. The discharge consists of the cathode and anode sheathes and a negative glow plasma without a positive column. The plasma is created by the energetic electrons emitted by the cathode and accelerated by the near-cathode sheath to the energies above the ionization potential for the gas atoms. Inelastic collisions of the energetic electrons with atoms create slow thermal electrons and ions. A typical dimension of the discharge, L, is less than electron energy relaxation length, λ_ε, which for the noble gases and electrons with energies in elastic collision range (electron energy is below the first excitation potential), is typically on the order of 10/p·cm, where p is the gas pressure in Torr [2, 3]. For atmospheric pressures, it is found that λ_ε 100 μm and $L < \lambda_\varepsilon$ for typical microdischarges. The short dc discharges, including microdischarges, can be used, for instance, for plasma-chemical and surface modification applications [7], as well as, for light sources [8], analytical sensors [9, 10], and plasma electronic devices [11]. In contrast to semiconductor devices, plasma discharges can be used under harsh conditions related, for example, to high temperatures and radiation levels of damaged nuclear plants (such as Fukushima-like disasters).

It is known (see, for example, refs. 12–17) that dc electric discharges can be unstable with respect to excitation of various types of oscillations and instabilities. Oscillations are excited due to ionization plasma instability related to the falling volt-ampere characteristic of the discharge or part of it (see refs. 12 and 18). These oscillations are affected by external electric circuits. There are many regimes of oscillations, which can be identified. While plasma instabilities, in principle, can be harnessed for some purposes, for example, for the generation of voltage oscillations, they are harmful for many other applications, for instance, for the development of current and voltage stabilization devices, and, in this case, they should be suppressed. Mitigation of these oscillations may not be a very simple task to accomplish. In this paper, we demonstrate how it is possible to achieve reliable suppression of the discharge oscillations by making use of an external auxiliary electrode. The method works for short dc discharge with thermal emission cathode in the nonlocal pressure regime $L < \lambda_\varepsilon \approx 10/p\cdot cm$.

An additional electron loss caused by the external electrode changes the voltage-ampere characteristic from falling to increasing with current. That makes the discharge stable and suppresses onset of the oscillations. It was experimentally demonstrated that a placement of the auxiliary electrode outside the discharge and subtraction of the anode current through a small opening in the anode provided the desirable effect.

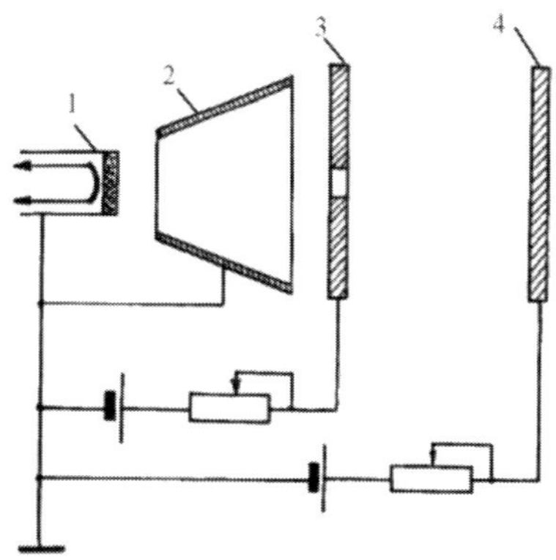

Figure 1. The scheme of the experimental device. Discharge takes place between cathode (1) and anode with a small opening (3) and is restricted radially by a cone screen (2). Auxiliary electrode (4) can provide the discharge plasma stabilization.

The experimental device for manifestation of the above effect is shown schematically in Figure 1. The plasma was created by the discharge, which existed between a grounded heated cathode and a positively biased anode. The cathode was a disc with a diameter of 1.0 cm. The molybdenum anode with an external diameter of 3.0 cm had a thickness of 0.2 cm and a central opening with an internal diameter of 0.2 cm. The molybdenum auxiliary electrode had a diameter of 3.0 cm and was placed near the outer part of the anode. The distance between the cathode and anode was 0.8 cm, and the distance between the anode and the auxiliary electrode could be changed from 0.1 to 5 cm. A conical electrode (screen) restricted the discharge plasma in the radial direction. The conical screen was electrically connected to the cathode and, therefore, it was also grounded. Electrons emitted from the cathode were accelerated by the cathode fall and moved towards the anode. Because the cross sectional area of the opening was small compared to the anode cross sectional area, the energetic electrons were lost mostly on the anode and not in the opening. A cylindrical movable tantalum electric probe (not shown in Figure 1) has been used to measure the EEDF, plasma potential, and plasma density [19, 20]. The probe had a diameter of 0.07 mm and length of 1 mm. It was introduced into the plasma, perpendicular to the axis of the device, for making measurements in the interelectrode gap. It could also be used for the measurements near the opening. The measurements were conducted in spectrally pure helium at pressures from 1 to 10 Torr with a discharge current up to 5 A.

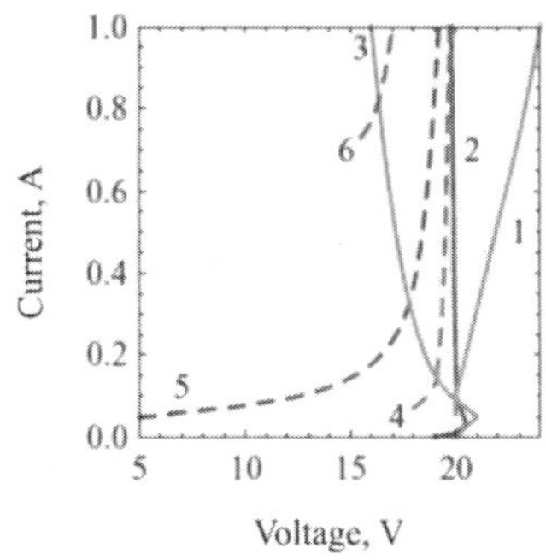

Figure 2. The discharge current-voltage characteristics for helium gas pressure of 0.6 Torr (1), 1 Torr (2), and 4 Torr (3). The diode regime is shown by solid curves (no current to the auxiliary electrode). Regimes with the auxiliary electrode are shown by the dashed curves for helium gas pressure of 1 Torr and different currents to the auxiliary electrode (0.1 A (4) and 0.4 A (5)). Curve (6) is for helium gas pressure of 4 Torr and current to the auxiliary electrode of 0.1 A.

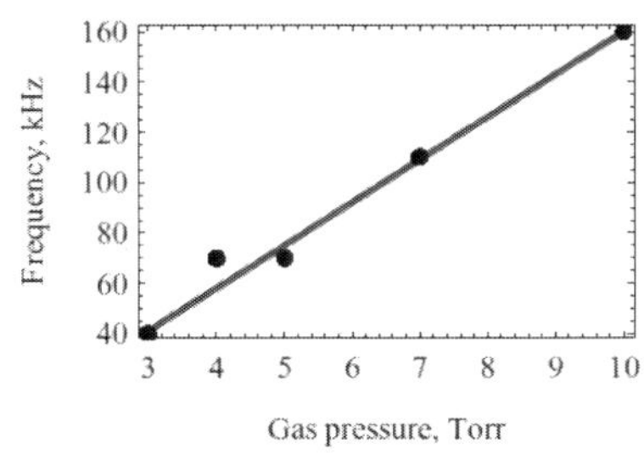

Figure 3. Characteristic frequencies of the discharge voltage oscillations with respect to the gas pressures in helium discharge.

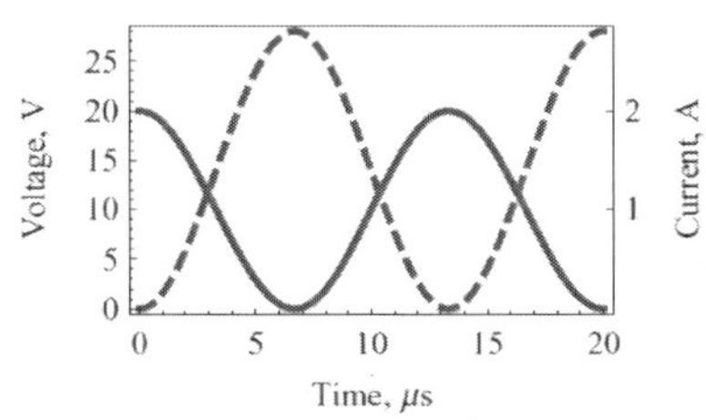

Figure 4. Typical voltage (dashed curve) and current (solid curve) oscillations. Gas pressure is 5 Torr.

In Figure 2, the typical current-voltage experimental characteristics of the cathode-anode gap are presented. Curves 1, 2, and 3 correspond to the case without collection of the current by the auxiliary electrode (diode regime). It is possible to see from Figure 2 that in the diode regime, the discharge has a positive discharge differential resistance for the pressures below 1 Torr (curve 1). Increasing the gas pressure to 1 Torr leads to transformation of the discharge differential resistance to the slightly negative one (curve 2). Further increase of the gas pressure yields higher negative discharge differential resistance of the discharge (curve 3). The presence of the negative discharge differential resistance may lead to plasma instabilities and oscillations of the discharge voltage and current. For gas pressures between 1 and 2 Torr when the discharge differential resistance was negative, voltage oscillations have wide-band spectra and amplitude on the order of 1 V. With the increase of the gas pressure, the oscillations transform into the narrow-band spectrum type. The characteristic frequencies of those oscillations are almost linear functions of the pressure, as shown in Figure 3. The frequencies weakly depend on the discharge current. The amplitude of oscillation can reach 30 V with a 100% modulation of the discharge current and voltage as it is shown in Figure 4. Further increase of the gas pressure, higher than 10 Torr, leads to transformation of the plasma with nonlocal EEDF to plasma with local EEDF ($L > \lambda_\varepsilon \approx 10/p\cdot cm$) and that could lead to the complete disappearance of the oscillations.

The nature of the oscillation modifications with discharge parameters can be clarified by the differences in the processes of the plasma production and loss. The main processes of the plasma production are direct and stepwise ionizations of the helium atoms. Under the studied conditions, the charged particle loss is mainly due to their diffusion to the walls or electrodes, and then recombination on the walls. Slow, thermal electrons cannot reach the walls and cathode (due to the high negative potentials) and can move to the anode only. Their current to the anode is restricted by the near-anode potential drop. In the limit of zero near-anode potential drop, the electron current of slow electrons is given by

$$I_{ch} = \frac{1}{4} e N_e V_T S, \tag{1}$$

where e is the electron charge, N_e is the electron density near the anode, V_T is the thermal speed of electrons, and S is the area of anode. For example, for the helium gas pressure of 5 Torr, N_e = 10^{11} cm^{-3} and T_e = 1 eV, eq. (1) gives $I_{ch} \approx 1.7$ A, whereas the experimental value is I_m = 2 A (see Figure 4). The estimated ion drift time in the discharge gap is an order of several milliseconds, which is in a good agreement with frequencies of the oscillations. Disappearance of the oscillations for the gas pressures higher than 10 Torr suggest that the oscillations are the feature of the nonlocal regime only ($L > \lambda_\varepsilon \approx 10/p\cdot cm$).

The collection of the anode current to the auxiliary electrode leads to a transformation of the discharge differential resistance from the negative to the positive one, as evident in Figure 2 from curves 4, 5, and 6. Curves 4 and 5 were obtained for the helium gas pressure of 1 Torr and curve 6 is for 4 Torr. Each curve was obtained with fixed auxiliary electrode current. The cathode current is the sum of the anode current and the auxiliary electrode current. The experimental data show that an increase in the current drawn to the auxiliary electrode makes discharge differential resistance even more positive (compare curves 4 and 5 in Figure 2). In all cases with the positive differential resistance, the oscillations are suppressed and are practically absent.

Thus, small openings in boundaries of nonlocal plasma allow the subtraction of slow electrons while practically not affecting energetic plasma electrons. This allows transformation of the negative discharge differential resistance to positive and effective suppression of plasma oscillations and instabilities.

REFERENCES

[1] Lieberman M. A., A. J. *Lichtenberg Principles of Plasma Discharge and Material Processing* (Wiley, New York, 2005).
[2] Tsendin L. D., *Phys.-Uspekhi* 53, 139 (2010).
[3] Kaganovich D., V. I. Demidov, S. F. Adams, and Y. Raitses, *Plasma Phys. Controlled Fusion* 51, 124003 (2009).
[4] DeJoseph, C. A. Jr., V. I. Demidov, and A. A. Kudryavtsev, *Phys. Plasmas* 14, 057101 (2007); *IEEE Trans. Plasma Sci.* 34, 825 (2006).
[5] Demidov V. I., C. A. DeJoseph, Jr., and A. A. Kudryavtsev, *Phys. Rev. Lett.* 95, 215002 (2005).
[6] Kolobov V. I., L. D. Tsendin, *Phys. Rev.A* 46, 7837 (1992).
[7] Al-Bataineh S. A. et al., *Plasma Processes Polym.* 9, 638 (2012).
[8] Higashiguchi T. et al., *J. Appl. Phys.* 109, 013301 (2011).
[9] Eun C. K. and Y. B. Gianchandani, *IEEE J. Quantum Electron.* 48(6), 814 (2012).
[10] Demidov V. I., S. F. Adams, J. Blessington, M. E. Koepke, and J. M. Williamson, *Contrib. Plasma Phys.* 50, 808 (2010).
[11] Chen K. F. and J. G. Eden, *Appl. Phys. Lett.* 93, 161501 (2008).
[12] Raizer Y. P., *Gas Discharge Physics* (Springer, New York, 1997).
[13] Petrovic Z. Lj., I. Stefanovic, S. Vrhovac, and J. Zivkovic, *J. Phys.* IV 7, C4-341 (1997).
[14] Phelps V. et al., *Phys. Rev.E* 47, 2825 (1993).
[15] Stefanovich I. et al., *J. Appl. Phys.* 110, 083310 (2011).
[16] Greiner F., T. Klinger, A. Rohde, and A. Piel, *Phys. Plasmas* 2, 1822 (1995).
[17] Greiner F., T. Klinger, and A. Piel, *Phys. Plasmas* 2, 1810 (1995).
[18] Fedotov M. A., I. Kaganovich, and L. D. Tsendin, *Sov. Phys. Tech. Phys.* 39, 241 (1994).
[19] Godyak V. A. and V. I. Demidov, *J. Phys. D: Appl. Phys.* 44, 233001 (2011).
[20] Demidov V. I., S. V. Ratynskaia, and K. Rypdal, *Rev. Sci. Instrum.* 73, 3109 (2002).

In: Applied Aspects of Nano-Physics ...
Editors: Kirill Levine and Andrey Syrkov

ISBN: 978-1-53614-708-7

Chapter 46

SPECIAL FEATURES OF THE THERMOELECTRIC MATERIALS BASED ON ANTIMONY AND BISMUTH TELLURIDES PRODUCED BY DIRECTIONAL CRYSTALLIZATION

Sergey A. Nemov[1,2], Arseny A. Rulimov[1,*]
and Alexandr V. Shchegol'kov[3]
[1]Peter the Great Saint Petersburg Polytechnic University,
Saint Petersburg, Russia
[2]Saint Petersburg Electrotechnical Univeristy, Saint Petersburg, Russia
[3]Tambov State Technical University, Tambov, Russia

ABSTRACT

Attempts to improve the properties of $Bi_xSb_{2-x}Te_3$ solid solution produced by directional crystallization are still going on. A lot of attention is paid to the selection of the optimal technology. It is argued that selenium addition increases the thermoelectric properties of $Bi_xSb_{2-x}Te_3$ samples by reducing the number of defects in Sb_2Te_3 crystal lattice. Besides we confirm the heterogeneity of the thermoelectric coefficient in $Bi_xSb_{2-x}Te_3$ samples' cross section and show annealing effect on this heterogeneity.

Keywords: $Bi_xSb_{2-x}Te_3$, Vertical zone melting, Heterogeneity of the thermoelectric properties, Annealing, Selenium

---

* Corresponding Author's Email: rulimmmov@gmail.com

INTRODUCTION

Production of low-temperature thermoelectric materials based on Sb and Bi tellurides remains an all-important area for the entire thermoelectric industry. Despite the long-standing reputation of these compounds, nowadays the theoretical base of the enterprises is being actively updated with the information of such solid solutions like $Bi_xSb_{2-x}Te_3$. The main problem in the production of corresponding materials is the high sensitivity to technological parameters, which unjustified variation entails loss of both thermoelectric and mechanical properties. Therefore, so much attention is paid to the selection of optimal technological parameters.

MATERIALS AND METHODS

$Bi_xSb_{2-x}Te_3$ solid solution with antimony telluride content of 70 to 80 mol. % is widely used in thermal converters with a working temperature range from 300K to 600K. Such a choice of composition is explained by the fact that from the point of view of the performance factor for devices based on $A_2^V B_3^{VI}$ materials the hole concentration $p \approx 1 \cdot 10^{19}$ cm^{-3} is recognized as optimal [1]. Sb_2Te_3 has the hole concentration $p \approx 1 \cdot 10^{20}$ cm^{-3} and for Bi_2Te_3 this value is equal to several units multiplied by 10^{18} cm^{-3}.

As we know in accordance with the Ioffe idea in $Bi_xSb_{2-x}Te_3$ system the lattice component of the thermal conductivity (χ_{ph}) is significantly lower in comparison with the χ_{ph} values of the individual components (bismuth and antimony tellurides). For the above-mentioned composition range the χ_{ph} value is minimal for the whole solid solution series [2].

Directional crystallization methods have established themselves as basic for the production of thermoelectric materials including the above-mentioned composition. *Growth rate anisotropy* (the growth rate in the direction across the cleavage plane is much less than along it) makes it possible to grow crystals with a directed structure, in which the value of the thermoelectric efficiency ZT is maximized along the growth direction [3].

EXPERIMENTAL AND DISCUSSION

During the experiments we have investigated $Bi_xSb_{2-x}Te_3$ samples with the hole conductivity to detect possible heterogeneity of thermoelectric properties. The samples had been produced by vertical zone melting. Particularly we have attempted to plot a distribution of the thermoelectric coefficient in the samples' cross section using thermal measuring probe. Typical distributions are shown in Figure 1.

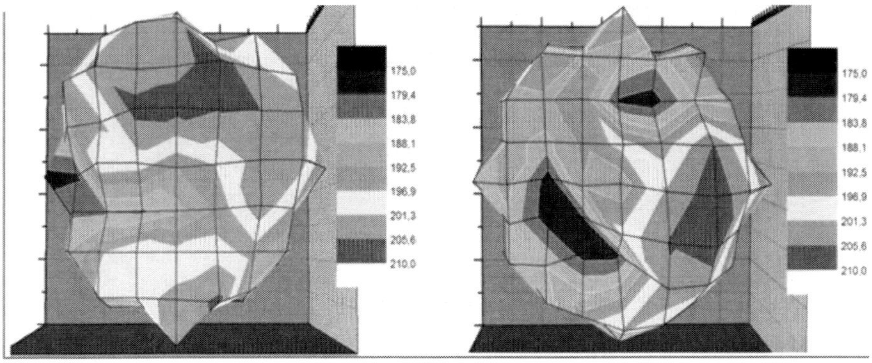

Figure 1. The distributions of the thermoelectric coefficient [μV/K] in two $Bi_xSb_{2-x}Te_3$ samples' cross section.

The probe's imprecision due to differences in the surface condition and in tip clamping was established empirically: its value is equal to one percent. Thus the thermal probe measurements are completely suitable for qualitative assessment of the distributions of the thermoelectric coefficient.

The thermoelectric coefficient, or the Seebeck coefficient, α is an extremely important thermoelectric parameter included in the thermoelectric efficiency formula to the second power:

$$Z = \frac{\alpha^2 \cdot \sigma}{\chi}, \tag{1}$$

where:

Z – the thermoelectric efficiency (determine the performance factor);
σ – the conductivity;
χ – the thermal conductivity.

The next step was the analysis of annealing effect on the above distributions. The dissolution of excess tellurium is known to be the main purpose of annealing such p-type thermoelectric materials. The diffusion rate of tellurium into the crystal lattice decreases with increasing temperature, so annealing is often carried out at a temperature T ≤ 250 °C. Results of annealing are shown in Figure 2.

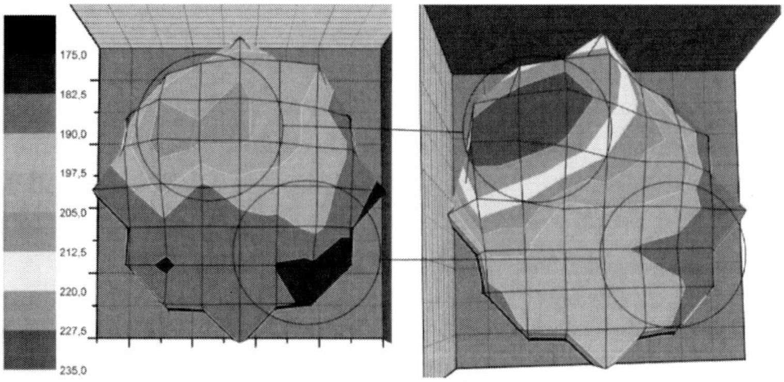

Figure 2. Annealing effect on the distribution of the thermoelectric coefficient [μV/K].

For precise adjustment of the composition and, as a consequence, useful properties of the samples and final products, it is expedient to introduce selenium into the composition, i.e., create quaternary $(Bi_xSb_{2-x})(Te_{3-y}Se_y)$ solid solution. The fact is that selenium additive smoothly reduces the hole concentration and the number of defects in Sb_2Te_3 crystal lattice (see Table 1 and Figure 3), which causes an increase in the hole mobility. All of the above have a positive effect on the values of σ and α [4, 5].

Table 1. The parameters of the valence band in p-$Sb_2Te_{3-x}Se_x$ at a temperature of 100 K. m_0 is the free electron mass; m_{d1} and m_{d2} are the density-of-state effective masses; ΔE_v is the energy gap between non-equivalent extremes of the valence band

	$Sb_2Te_{2.99}Se_{0.01}$	$Sb_2Te_{2.95}Se_{0.05}$	$Sb_2Te_{2.9}Se_{0.1}$
$p\cdot 10^{19}$, cm^{-3}	8.5	8.4	8.2
m_{d1}/m_0	0.6	0.6	0.5
m_{d2}/m_0	1.8	1.5	1.4
ΔE_v, eV	0.125	0.13	0.14

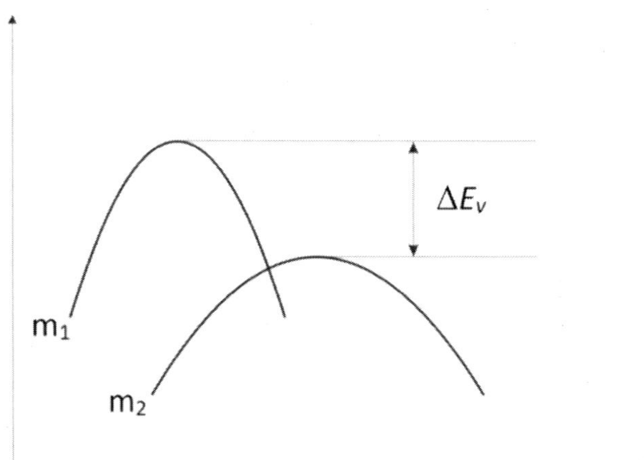

Figure 3. Qualitative form of the valence band in p-$Sb_2Te_{3-x}Se_x$.

It also should be noted that thermoelectric materials based on Sb and Bi tellurides (especially with the hole conductivity) have poor mechanical strength. In this context the appending of carbon nanomaterials into the composition looks promising. Balance distribution of carbon nanomaterials in $Bi_xSb_{2-x}Te_3$ can probably solve a problem with mechanical properties.

RESULTS AND CONCLUSIONS

During the research we have confirmed and described the heterogeneity of thermoelectric properties in $Bi_xSb_{2-x}Te_3$ samples produced by vertical zone melting. Consistent patterns in the distribution of the thermoelectric coefficient along samples' cross section were not found. This is explained by the presence of compositional fluctuations at the crystallization front.

The dissolution of excess tellurium during the annealing process increased the value of the thermoelectric coefficient in each cross section point and also increased the difference between the maximum and minimum values.

Small selenium addition to the charge improves the thermoelectric properties of $Bi_xSb_{2-x}Te_3$ samples by reducing the number of defects in Sb_2Te_3 crystal lattice.

All of the above results and suppositions will allow optimizing the production technology of low-temperature thermoelectric materials based on Sb and Bi tellurides.

REFERENCES

[1] Gol'tsman, B.M., Dashevskii, Z.M., Kaidanov, V.I., Kolomoets, N.V. (1985). *Film Thermoelements: Physics and Application*. Nauka, Moscow.

[2] Gol'tsman, B.M., Kudinov, V.A., Smirnov, I.A. (1972). *Semiconductor thermoelectric materials based on* Bi_2Te_3. Nauka, Moscow: 91.

[3] Lavrent'ev, M. G., Osvensky, V. B., Pivovarov, G. I., Sorokin, A. I., Karataev, V. V., Bublik, V. T., Tabachkova, N. Yu. (2015). Mechanical properties of bismuth and antimony solid solutions produced by directional crystallization and powder metallurgy. Thermoelectric materials and their applications. *Reports of the Interstate Conference*. The Ioffe Institute (Saint Petersburg).

[4] Ivanova, L. D., Sidorov, Yu. A. (1998). Electrical properties of Se-doped $Bi_{0.5}Sb_{1.5}Te_3$ crystals. *Inorganic Materials,* 34 (3): 222-224.

[5] Luk'yanova, L. N., Kutasov, V. A., Popov, V. V., Konstantinov, P. P. (2004). Galvanomagnetic and thermoelectric properties of p-$Bi_{2-x}Sb_xTe_{3-y}Se_y$ solid solutions at low temperatures (< 220 K). *Physics of the Solid State,* 46 (8): 1404-1409.

In: Applied Aspects of Nano-Physics …
Editors: Kirill Levine and Andrey Syrkov
ISBN: 978-1-53614-708-7

Chapter 47

SYNTHESIS, IDENTIFICATION AND PHYSICAL-CHEMICAL PROPERTIES OF BIS-ADDUCT OF FULERENE C_{60} WITH HISTIDINE

***Vlada V. Petrenko*[1], *Nikolay A. Charykov*[1,2], *Konstantin N. Semenov*[1,3], *Viktor A. Keskinov*[1,*], *Aleksey V. Kurilenko*[1], *Irina V. Semenyuk*[1] and *Nikolay M. Safiannikov*[2]**

[1]Saint-Petersburg State Technological Institute (Technical University), Saint-Petersburg, Russia
[2]Saint-Petersburg State Electro-technical University (LETI), Saint-Petersburg, Russia
[3]Saint-Petersburg State University, Saint-Petersburg, Russia

ABSTRACT

The heterogeneous – non-catalytic method of the synthesis and purification (by water – methanol re-crystallization) of bis-adduct of fullerene C_{60} with amino-acid histidine is described. Identification of bis-adduct was provided by the following methods of physical-chemical analysis: C-H-N-O Element analysis, High performance liquid chromatography, IR-spectroscopy, Electronic spectroscopy. Volume properties of bis-

---

* Corresponding Author's Email: keskinov@mail.ru.

adduct of fullerene C_{60} with histidine water solutions - $C_{60}(C_6N_3H_8O_2)_2(H)_2 – H_2O$ at 298 K (concentration dependencies of density, average and partial molar volumes of both components) were determined with the help of quartz pycnometers. Concentration dependencies of refraction properties (refraction indexes, specific and molar refractions of the solutions and bis-adducts) at 298 K were determined with the help of refractometers. Using method of isothermal saturation in ampules, poly-thermal solubility of $C_{60}(C_6N_3H_8O_2)_2(H)_2$ in water solutions was determined in the temperature range 293 – 353 K. By the method of visible light scattering, concentration dependencies of linear dimensions of nano-cluster associates and electro-kinetic ζ – potentials were determined at 298 K. By the method of cryo-metrywith the help of Beckmann thermometer concentration dependence of the decrease of liquidus temperatures was determined at the temperatures nearby 273 K. On the base of last data excess function of both components - $C_{60}(C_6N_3H_8O_2)_2(H)_2$ and H_2O (activities and activity coefficients) were calculated, using novel semi-empirical Virial Decomposition Asymmetric Model (VD-AS). Also with the help of VD-AS diffusional instability concentration region was calculated.

Keywords: bis-adduct, fullerene C_{60}, histidine, synthesis, heterogeneous – non-catalytic method, identification, physical-chemical analysis, water solutions, physical-chemical properties

INTRODUCTION

Water soluble derivatives of fullerenes is a perspective class of compounds due to possibilities of their application in various fields of science and technology especially in biology and medicine due to well defined membranotrophic, cytoprotective, radioprotective, antioxidant, antimicrobial, antiviral and transport properties [1-3]. This chapter is devoted to the development of the cycle of works, devoted to the synthesis, identification and the investigation pf physical-chemical properties of the adducts of light fullerenes (C_{60} or C_{70}) and amino-acids: alanine, lysine, histidine, arginine, glycine, proline etc. Stromet et al. prepared fullerene amino acids by dipolar addition to C60of either the Boc- or Fmoc-Na-protected azido amino acids derived from phenylalanine and lysine [4]. It was determined that prepared aminoacids are a mixture of 5,6-open (major product) and 6,6-closed (minor product) derivatives. Kotelnikova et al. synthesized the water-soluble C60 derivatives with DL-alanine and investigated the influence of the obtained compounds on structure and permeability of the lipid bilayer of phosphatidylcholine liposomes [5]. Hu et al. synthesized the C60 derivative with β-alanine, cystine and arginine [6]. Kumar et al. synthesized and studied DNA cleavage efficiency upon visible light (in the presence of nicotinamide adenine dinucleotide) ofthe C60-lysine derivative [7]. Jiang et al. synthesized novel water-soluble C60-glycine derivative [8]. The results of cytotoxicity assay of cancer cell lines showed thatC60-glycine derivative in a dose-dependent manner increases cell death. Authors of [9, 10] synthesized hybrid structures based on fullereneC60 with attached proline amino acid (methyl ether of N-[(β-alanylhistidyl-ethyl) fullerenyl] proline, methyl ether of N-[(nitroxyethyl)fullerenyl] proline, methyl ether of N-[(2′,3′-dinitroxy-propyl)fullerenyl] proline, methyl ether of N-[mono-hydro-fullerenyl] proline,carnosine). It was determined that all studied compounds possess anantioxidant activity and inhibited glutamate induced Ca^{2+} uptake into synap to some of the rat brain cortex. Some articles were fulfilled with the participation of the authors of the presented chapter [11-17].

MATERIALS AND METHODS

Yield of bis-adduct was 76% from the theoretical one.

The heterogeneous – non-catalytic method of the synthesis of bis-adduct of fullerene C_{60} with amino-acid histidine -$C_{60}(C_6H_8N_3O_2)_2(H_2)$ or 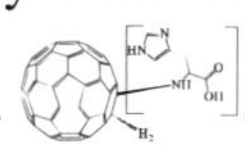 is described. Histidine (mass m = 3.2g), base NaOH (m = 15.8g) were dissolved in 54 cm^3 of H_2O; 130 cm^3 of the solution of fullerene C_{60} in o-xylene with the concentration 7.62g/$dm^3$was prepared by direct dissolution. Then ethanol (C_2H_5OH) (m = 267g) was added to the both "o-xylene" and "water" phases and the heterogeneous mixture was stirred for 7 days. Then phases were divided, "water" phase was evaporated and bis-adduct $C_{60}(C_6H_8N_3O_2)_2(H_2)$ was precipitated by methanol (CH_3OH). Purification of bis-adduct was provided by triple recrystallization from water solutions by methanol. Additionally bis-adduct $C_{60}(C_6H_8N_3O_2)_2(H_2)$ was washed from the impurities of sodium-contain forms by HCl acid in Soxhlet apparatus.

IDENTIFICATION

Identification of bis-adduct $C_{60}(C_6H_8N_3O_2)_2(H_2)$ was provided by the following methods of physical-chemical analysis:

C-H-N-O Element analysis (CHNS/O Analyzer 2400 Series II) proved formula of crystal hydrate $C_{60}(C_6N_3H_8O_2)_2(H)_2*24H_2O$;

High performance liquid chromatography (HPLC PerkinElmer) confirmed purity of bis-adduct 96.5 mass %;

IR-spectroscopy (IRTracer-100, Shimadzu) in wavenumbers: ν = 400 – 4000cm^{-1} demonstrates absorption characteristic peaks for: valent O-H (3392cm^{-1}), C-H, CH_2 (2937, 2863 cm^{-1}) C = O (1070cm^{-1}), deformation O-H (1405cm^{-1}), N-H (1593cm^{-1}) oscillations and long-wave characteristic oscillations of C_{60} fullerene core (500 – 1000cm^{-1});

Electronic spectroscopy (UV-1800, Shimadzu), wavelength: λ = 190-1100 nm demonstrates no effects and monotonically increasing with wavelength decreasing light absorption. Also the feasibility of Bouguer-Lambert-Ber law at wavelength 330 nm was tested for the determination of $C_{60}(C_6N_3H_8O_2)_2(H)_2$ concentration in water solutions. The dependence of optical density on the solution concentration of bis-adduct concentration is represented in Figure 1.

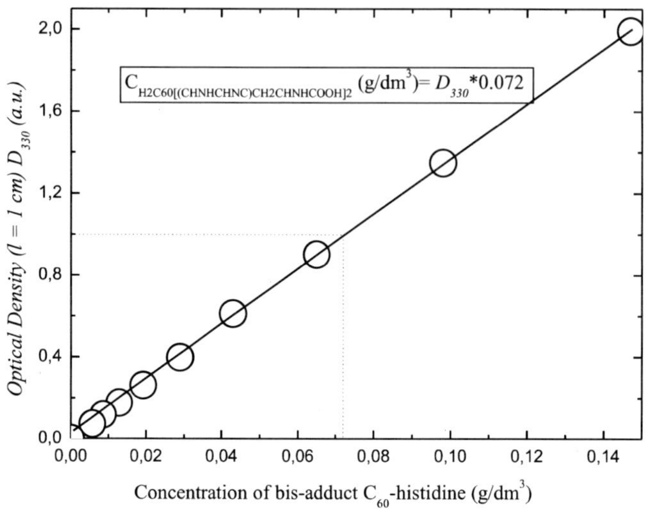

Figure 1. The feasibility of Bouguer-Lambert-Ber law at wavelength 330nm in water solution.

$$C_{C60(C6N3H8O2)2(H)2} = D_{330} * 0.072 \pm 0.02 \qquad (1)$$

where: $C_{C60(C6N3H8O2)2(H)2}$ is bis-adduct concentration in g/dm^3, D_{330} is solution optical density at wavelength λ=330 nm and length of optical path l = 1cm.

Experimental Section

Volume properties of bis-adduct of fullerene C_{60} with histidine water solutions - $C_{60}(C_6N_3H_8O_2)_2(H)_2 - H_2O$ at 298 K (concentration dependencies of density, average and partial molar volumes of both components) were determined with the help of quartz pycnometers. Concentration dependencies of refraction properties (refraction indexes, specific and molar refractions of the solutions and bis-adducts) at 298 K were determined with the help of refractometers.

Using method of isothermal saturation in ampules, poly-thermal solubility of $C_{60}(C_6N_3H_8O_2)_2(H)_2$ in water solutions was determined in the temperature range 293 – 353 K. The diagram of solubility is represented in Figure 2. Onecan see, that it consists of two branches of crystallization of crystal hydrate $C_{60}(C_6N_3H_8O_2)_2(H)_2*24H_2O$ and $C_{60}(C_6N_3H_8O_2)_2(H)_2$.

By the method of visible light scattering, concentration dependencies of linear dimensions of nano-cluster associates and electro-kinetic ζ – potentials were determined at 298 K.

By the method of cryo-metry with the help of Beckmann thermometer concentration dependence of the decrease of liquidus temperatures was determined at the temperatures nearby 273 K (see Figure 3). On the base of last data excess function of both components - $C_{60}(C_6N_3H_8O_2)_2(H)_2$ and H_2O (activities and activity coefficients) were calculated, using novel semi-empirical Virial Decomposition Asymmetric Model (VD-AS). Also with the help of VD-AS diffusional instability concentration region was calculated.

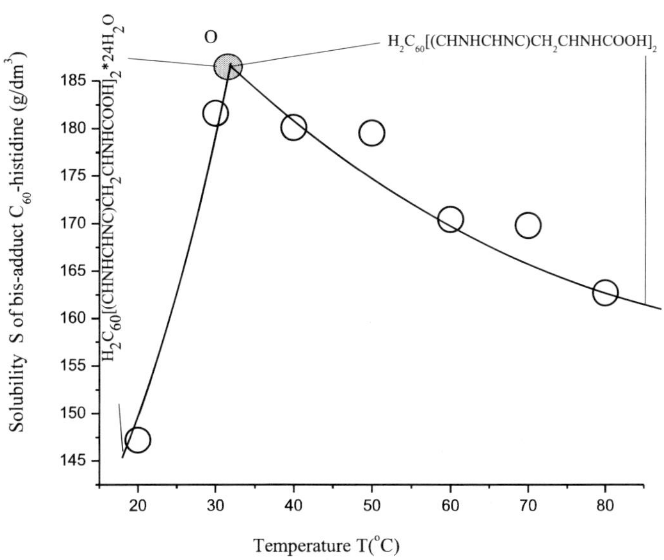

Figure 2. Solubility of $C_{60}(C_6N_3H_8O_2)_2(H)_2$in H_2O in temperature range 20-80°C.

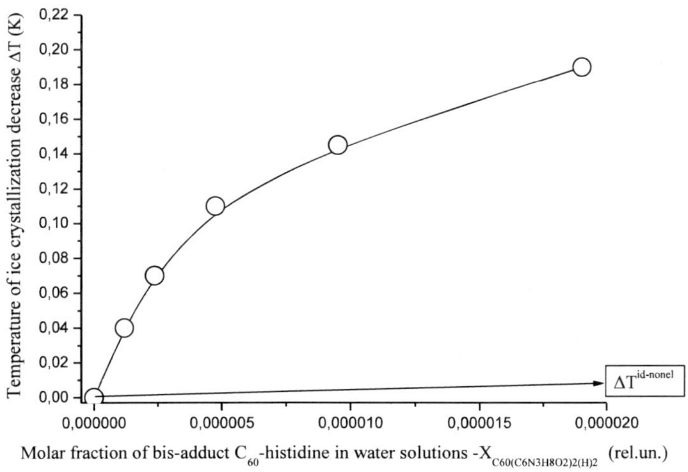

Figure 3. Decrease of liquidus temperatures depending on molar fraction of $C_{60}(C_6N_3H_8O_2)_2(H)_2$in H_2O solution at 272.97 - 273.15 K. Arrow symbolizes ideal non-electrolyte solution.

CONCLUSION

Thus, bis-adduct of fullerene C_{60} with histidine was synthesized, identified by modern methods of physical-chemical analysis, main physical-chemical properties of its water solutions were investigated.

ACKNOWLEDGMENT

This work was supported by Russian Found of Fundamental Investigations – RFFI (Projects no. 16-08-01206, 18-08-00143 A), by the Grant of the President of Russian Federation for supporting young scientists MK- 4657.2015.3. Research was performed with using the equipment of the Resource Centers 'GeoModel', Center for Chemical Analysis and Materials Research and Center for Thermogravimetric and Calorimetric Research of Research park of Saint Petersburg State University.

REFERENCES

[1] F. Cataldo, T. Da Ros, Carbon Materials: Chemistry and Physics: *Medical Chemistry and Pharmacological Potential of Fullerenes and Carbon Nanotube*, Springer, 2008.

[2] L. N. Sidorov, M. A. Yurovskaya, *Fullerenes, Ekzamen*, Moscow, 2005.

[3] L. B. Piotrovskii, O. I. Kiselev, *Fullerenes in Biology*, 2006 (Rostok, Saint-Petersburg).

[4] T. A. Strom, A. R. Barron, *Chem. Commun.*, 46 (2010) 4764 - 4766.

[5] R. A. Kotelnikova, A. I. Kotelnikov, G. N. Bogdanov, V. S. Romanova, E. F. Kuleshova, Z. N. Parnes, M. E. Volpin, *FEBS Lett.,* 389 (1996) 111 - 114.

[6] Z. Hu, W. Guan, W. Wang, L. Huang, X. Tang, H. Xu, Z. Zhu, X. Xie, H. Xing, *Carbon*, 46. (2008) 99 - 109.

[7] A. Kumar, M. V. Rao, S. K. Menon, *Tetrahedron Lett.,* 50 (2009) 6526 - 6530.

[8] G. Jiang, F. Yin, J. Duan, G. Li, *J. Mater. Sci. Mater. Med.,* 26 (2015) 1 - 7.

[9] V. V. Grigoriev, L. N. Petrova, T. A. Ivanova, R. A. Kotelnikova, G. N. Bogdanov, D. A., Poletayeva, I. I. Faingold, D. V. Mishchenko, V. S. Romanova, A. I. Kotel'nikov, S. O. Bachurin, *Biol. Bull.*, 38 (2011) 125 - 131.

[10] L. V. Tatyanenko, O. V. Dobrokhotova, R. A. Kotelnikova, D. A. Poletayeva, D. V., Mishchenko, I. Y. Pikhteleva, G. N. Bogdanov, V. S. Romanova, A. I. Kotelnikov, *Pharm. Chem. J.,* 45 (2011) 329 - 332.

[11] M. Y. Matuzenko, D. P. Tyurin, O. S. Manyakina, K. N. Semenov, N. A. Charykov, K. V. Ivanova, V. A. Keskinov, *Nanosystems: Phys. Chem. Math.*, 6 (2015) 704-714.

[12] M. Y. Matuzenko, A. A. Shestopalova, K. N. Semenov, N. A. Charykov, V. A. Keskinov, *Nanosystems: Phys. Chem. Math.,* 6 (2015) 715 - 725.

[13] A. A. Shestopalova, K. N. Semenov, N. A. Charykov, V. N. Postnov, N. M. Ivanova, V. V. Sharoyko, V. A. Keskinov, D. G. Letenko, V. A. Nikitin, V. V. Klepikov, I. V. Murin, *J. Mol. Liq.,* 211 (2015) 301- 307.

[14] O. S. Manyakina, K. N. Semenov, N. A. Charykov, N. M. Ivanova, V. A. Keskinov, V. V., Sharoyko, D. G. Letenko, V. A. Nikitin, V. V. Klepikov, I. V. Murin, *J. Mol. Liq.,* 211. (2015) 487 - 493.

[15] K. N. Semenov, N. A. Charykov, I. V. Murin, Y. V. Pukharenko, *J. Mol. Liq.,* 201 (2015). 50 - 58.

[16] K. N. Semenov, N. A. Charykov, V. N. Keskinov, *J. Chem. Eng. Data,* 56 (2011) 230 - 239.

[17] K. N. Semenov, N. A. Charykov, I. V. Murin, Y. V. Pukharenko, *J. Mol. Liq.,* 201 (2015). 1 - 8.

In: Applied Aspects of Nano-Physics …
Editors: Kirill Levine and Andrey Syrkov
ISBN: 978-1-53614-708-7

Chapter 48

SYNTHESIS, IDENTIFICATION AND PROPERTIES OF TRIS-ADDUCT LIGHT FULLERENE C_{70} WITH INDISPENSABLE AMINO-ACID LYSINE

***Irina V. Semenyuk*[1], *Konstantin N. Semenov*[1,3], *Nikolay A. Charykov*[1,2], *Viktor A. Keskinov*[1,*], *Aleksey V. Kurilenko*[1], *Vlada V. Petrenko*[1] and *Nikolay M. Safiannikov*[2]**

[1]Saint-Petersburg State Technological Institute (Technical University), Russia
[2]Saint-Petersburg State Electro-technical University (LETI), Russia
[3]Saint-Petersburg State University, Russia

ABSTRACT

The heterogeneous – non-catalytic method of the synthesis and purification (by water – methanol re-crystallization) of tris-adduct of fullerene C_{70} with amino-acid lysine is described. Identification of bis-adduct was provided by the following methods of physical-chemical analysis: C-H-N-O Element analysis, High performance liquid chromatography, IR-spectroscopy, Electronic spectroscopy, Complex thermal analysis. Volume properties of tris-adduct of fullerene C_{70} with lysine water solutions - $C_{70}(C_6H_{13}N_2O_2)_3(H_3)$at 298 K (concentration dependencies of density, average and partial molar volumes of both components) were determined with the help of quartz pycnometers. Concentration dependencies of refraction properties (refraction indexes, specific and molar refractions of the solutions and tris-adducts) at 298 K were determined with the help of refractometers. By the method of visible light scattering, concentration

* Corresponding Author's Email: keskinov@mail.ru.

dependencies of linear dimensions of nano-cluster associates and electro-kinetic ζ – potentials were determined at 298 K. By the method of cryoscopy with the help of Beckmann thermometer concentration dependence of the decrease of liquidus temperatures was determined at the temperatures nearby 273 K. On the base of last data excess function of both components - $C_{70}(C_6H_{13}N_2O_2)_3(H_3)$ and H_2O (activities and activity coefficients) were calculated, using novel semi-empirical Virial Decomposition Asymmetric Model (VD-AS). Also with the help of VD-AS diffusional instability concentration region was calculated. Concentration dependencies of density, dynamic and kinematic viscosity at 293, 313 and 333 K was determined by the Stokes method of "falling gold ball" in water solutions of tris-adduct at 298 K. Concentration dependencies of hydrogen indicator (with the help of glass electrode), specific electro-conductivity (by RLC meter) were determined in tris-adduct water solutions at 298 K. On the base of both last data seeming dissociation degrees and concentration dissociation constants were calculated. Potential-metric titration of - $C_{70}(C_6H_{13}N_2O_2)_3(H_3)$ water solutions by acid (HCl) and base (NaOH) were provided.

Keywords: tris-adduct, fullerene C_{70}, lysine, synthesis, heterogeneous – non-catalytic method, identification, physical-chemical analysis, water solutions, physical-chemical properties

INTRODUCTION

Water soluble derivatives of fullerenes is a perspective class of compounds due to possibilities of their application in various fields of science and technology especially in biology and medicine due to well-defined membranotrophic, cytoprotective, radioprotective, antioxidant, antimicrobial, antiviral and transport properties [1–3]. This chapter is devoted to the development of the cycle of works, devoted to the synthesis, identification and the investigation pf physical-chemical properties of the adducts of light fullerenes (C_{60} or C_{70}) and amino-acids: alanine, lysine, histidine, arginine, glycine, proline etc. Stromet et al. prepared fullerene amino acids by dipolar addition to C60 of either the Boc- or Fmoc-Na-protected azido amino acids derived from phenylalanine and lysine [4]. It was determined that prepared amino acids are a mixture of 5,6-open (major product) and 6,6-closed (minor product) derivatives. Kotelnikova et al. synthesized the water-soluble C60 derivatives with DL-alanine and investigated the influence of the obtained compounds on structure and permeability of the lipid bilayer of phosphatidylcholine liposomes [5]. Hu et al. synthesized the C60 derivative with β-alanine, cystine and arginine [6]. Kumar et al. synthesized and studied DNA cleavage efficiency upon visible light (in the presence of nicotinamide adenine dinucleotide) of the C60-lysine derivative [7]. Jiang et al. synthesized novel water-soluble C60-glycine derivative [8]. The results of cytotoxicity assay of cancer cell lines showed that C60-glycine derivative in a dose-dependent manner increases cell death. Authors of [9, 10] synthesized hybrid structures based on fullerene C60 with attached proline amino acid (methyl ether of N-[(β-alanylhistidyl-ethyl)fullerenyl] proline, methyl ether of N-[(nitroxyethyl) fullerenyl] proline, methyl ether of N-[(2′,3′-dinitroxy-propyl)fullerenyl] proline, methyl ether of N-[mono-hydro-fullerenyl] proline, carnosine). It was determined that all studied compounds possess an antioxidant activity and inhibited glutamate induced Ca^{2+} uptake into synap to some of the rat brain cortex. Some articles were fulfilled with the participation of the authors of the presented chapter [11-17].

Materials and Methods

The heterogeneous – non-catalytic method of the synthesis of tris-adduct of fullerene C_{70} with amino-acid lysine - $C_{70}(C_6H_{13}N_2O_2)_3(H)_3$ is described. Histidine (mass m = 4.5 g), base NaOH (m = 15.8 g) were dissolved in 54 cm^3 of H_2O; 130 cm^3 of the solution of fullerene C_{70} in o-xylene with the concentration 8.9 $g/dm^3$was prepared by direct dissolution. Then ethanol (C_2H_5OH) (m = 250 g) was added to the both "o-xylene" and "water" phases and the heterogeneous mixture was stirred for 7 days. Then phases were divided, "water" phase was evaporated and tris-adduct $C_{70}(C_6H_{13}N_2O_2)_3(H)_3$ was precipitated by methanol (CH_3OH). Purification of tris-adduct was provided by triple recrystallization from water solutions by methanol. Additionally tris-adduct $C_{70}(C_6H_{13}N_2O_2)_3(H)_3$was washed from the impurities of sodium-contain forms by HCl acid in Soxhlet apparatus. Yield of tris-adduct was 79% from the theoretical one.

Experimental Section Identification

Identification of tris-adduct $C_{70}(C_6H_{13}N_2O_2)_3(H)_3$ was provided by the following methods of physical-chemical analysis:

C-H-N-O Element analysis (CHNS/O Analyzer 2400 Series II) proved formula of crystal hydrate $C_{70}(C_6H_{13}N_2O_2)_3(H)_3*18H_2O$;

High performance liquid chromatography (HPLC PerkinElmer) confirmed purity of tris-adduct 98.6 mass %;

IR-spectroscopy (IRTracer-100, Shimadzu) in wavenumbers: ν = 400 – 4000 cm^{-1} demonstrates absorption characteristic peaks for: valent O-H(3392 cm^{-1}), C-H(2937 cm^{-1}), CH_2(2863 cm^{-1}), C = O(1070 cm^{-1}), deformation O-H(1405 cm^{-1}), N-H(1593 cm^{-1}) oscillations and long-wave characteristic oscillations of C_{60} fullerene core (500 – 1000 cm^{-1}). IR spectrum, as an example is represented in Figure 1;

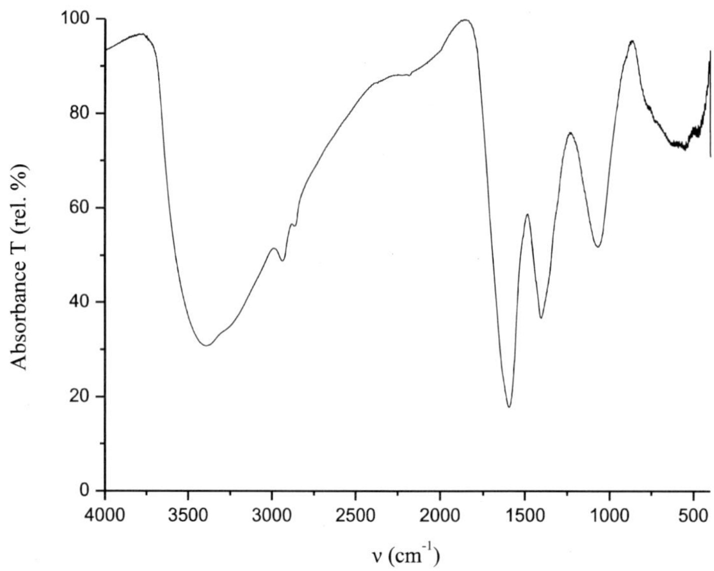

Figure 1. IR spectrum of $C_{70}(C_6H_{13}N_2O_2)_3(H)_3$.

Electronic spectroscopy (UV-1800, Shimadzu), wavelength: λ = 190-1100 nm demonstrates no effects and monotonically increasing with wavelength decreasing light absorption. Also the feasibility of Bouguer-Lambert-Ber law at wavelength 330 nm was tested for the determination of $C_{70}(C_6H_{13}N_2O_2)_3(H)_3$concentration in water solutions. One can see the validity of this law in all device optical range ($D_{330} = 0 \div 2.5$ – optical density at wavelength $\lambda = 330$ nm and light path l = 1 cm);

Complex thermal analysis shows stepwise external dehydration of crystal hydrates at 90 -210°C, stepwise decarboxylation, internal dehydration, denitrogenation at 300-810°C, C_{70} fullerene core oxidation at the temperatures higher than 900°C.

Physical-Chemical Properties

Volume properties of tris-adduct of fullerene C_{70} with lysine water solutions - $C_{70}(C_6H_{13}N_2O_2)_3(H_3)$ at 298 K (concentration dependencies of density, average and partial molar volumes of both components) were determined with the help of quartz pycnometers. Concentration dependence of partial molar volume of $C_{70}(C_6H_{13}N_2O_2)_3(H_3)$, as an example, is represented in Figure 2.

Concentration dependencies of refraction properties (refraction indexes, specific and molar refractions of the solutions and bis-adducts) at 298 K were determined with the help of refractometers.

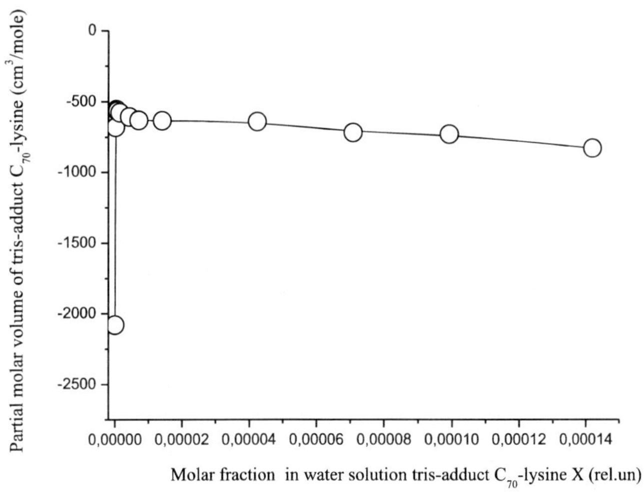

Figure 2. Concentration dependence of partial molar volume of $C_{70}(C_6H_{13}N_2O_2)_3(H_3)$.

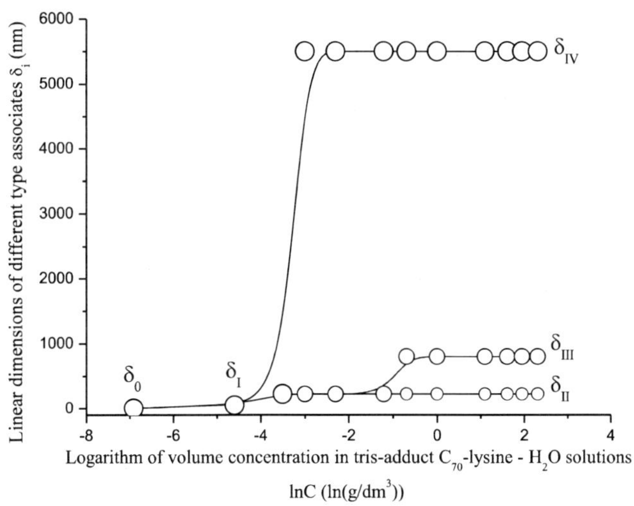

Figure 3. Concentration dependence of linear dimensions of different type associates - δ_i of $C_{70}(C_6H_{13}N_2O_2)_3(H_3)$ in semi-logarithmic scale.

By the method of visible light scattering, concentration dependencies of linear dimensions of nano-cluster associates (δ_i) and electro-kinetic ζ – potentials were determined at 298 K. Concentration dependence of $\delta_i$of $C_{70}(C_6H_{13}N_2O_2)_3(H_3)$ associates is represented in Figure 3.

By the method of cryo-metry with the help of Beckmann thermometer concentration dependence of the decrease of liquidus temperatures was determined at the temperatures nearby 273 K. On the base of last data excess function of both components - $C_{70}(C_6H_{13}N_2O_2)_3(H_3)$ and H_2O (activities and activity coefficients) were calculated, using novel semi-empirical Virial Decomposition Asymmetric Model (VD-AS). Also with the help of VD-AS diffusional instability concentration region was calculated.

Concentration dependencies of density, dynamic and kinematic viscosity at 293, 313 and 333 K was determined by the Stokes method of "falling gold ball" in water solutions of bis-adduct at 298 K. Concentration dependencies of hydrogen indicator (with the help of glass electrode), specific electro-conductivity (by RLC meter) were determined in bis-adduct water solutions at 298 K. On the base of both last data seeming dissociation degrees and concentration dissociation constants were calculated. Potential-metric titration of - $C_{70}(C_6H_{13}N_2O_2)_3(H_3)$ water solutions by acid (HCl) and base (NaOH) were provided.

CONCLUSION

Thus, tris-adduct of fullerene C_{70} with lysine was synthesized, identified by modern methods of physical-chemical analysis, main physical-chemical properties of its water solutions were investigated.

ACKNOWLEDGMENT

This work was supported by Russian Found of Fundamental Investigations – RFFI (Projects no. 16-08-01206, 18-08-00143 A), by the Grant of the President of Russian Federation for supporting young scientists MK- 4657.2015.3. Research was performed with using the equipment of the Resource Centers 'GeoModel,' Center for Chemical Analysis and Materials Research and Center for Thermogravimetric and Calorimetric Research of Research park of Saint Petersburg State University.

REFERENCES

[1] F. Cataldo, T. Da Ros, *Carbon Materials: Chemistry and Physics: Medical Chemistry and Pharmacological Potential of Fullerenes and Carbon Nanotube*, Springer, 2008.

[2] L. N. Sidorov, M. A. Yurovskaya, *Fullerenes*, Ekzamen, Moscow, 2005.

[3] L. B. Piotrovskii, O. I. Kiselev, *Fullerenes in Biology*, 2006 (Rostok, Saint-Petersburg).

[4] T. A. Strom, A. R. Barron, *Chem. Commun.* 46 (2010). 4764–4766.

[5] R. A. Kotelnikova, A. I. Kotelnikov, G. N. Bogdanov, V. S. Romanova, E. F. Kuleshova, Z. N. Parnes, M. E. Volpin, *FEBS Lett.* 389 (1996) 111–114.

[6] Z. Hu, W. Guan, W. Wang, L. Huang, X. Tang, H. Xu, Z. Zhu, X. Xie, H. Xing, *Carbon* 46. (2008) 99–109.

[7] Kumar, M. V. Rao, S. K. Menon, *Tetrahedron Lett.* 50 (2009) 6526–6530.

[8] G. Jiang, F. Yin, J. Duan, G. Li, *J. Mater. Sci. Mater. Med.* 26 (2015) 1–7.

[9] V. V. Grigoriev, L. N. Petrova, T. A. Ivanova, R. A. Kotelnikova, G.N. Bogdanov, D. A., Poletayeva, I. I. Faingold, D. V. Mishchenko, V.S. Romanova, A. I. Kotel'nikov, S. O. Bachurin, *Biol. Bull.* 38 (2011) 125–131.

[10] L. V. Tatyanenko, O. V. Dobrokhotova, R. A. Kotelnikova, D. A. Poletayeva, D.V., Mishchenko, I. Y. Pikhteleva, G. N. Bogdanov, V. S. Romanova, A. I. Kotelnikov, *Pharm. Chem. J.* 45 (2011) 329–332.

[11] M. Y. Matuzenko, D. P. Tyurin, O. S. Manyakina, K. N. Semenov, N. A. Charykov, K. V. Ivanova, V. A. Keskinov, Nanosystems*: Phys. Chem. Math.* 6 (2015) 704–714.

[12] M. Y. Matuzenko, A. A. Shestopalova, K. N. Semenov, N. A. Charykov, V. A. Keskinov, Nanosystems: *Phys. Chem. Math.* 6 (2015) 715–725.

[13] A. A. Shestopalova, K. N. Semenov, N. A. Charykov, V. N. Postnov, N. M. Ivanova, V. V. Sharoyko, V. A. Keskinov, D. G. Letenko, V. A. Nikitin, V. V. Klepikov, I. V. Murin, *J. Mol. Liq.* 211 (2015) 301–307.

[14] O. S. Manyakina, K. N. Semenov, N. A. Charykov, N. M. Ivanova, V. A. Keskinov, V. V., Sharoyko, D. G. Letenko, V. A. Nikitin, V. V. Klepikov, I.V. Murin, *J. Mol. Liq.* 211. (2015) 487–493.

[15] K. N. Semenov, N. A. Charykov, I. V. Murin, Y. V. Pukharenko, *J. Mol. Liq.* 201 (2015). 50–58.

[16] K. N. Semenov, N. A. Charykov, V. N. Keskinov, *J. Chem. Eng. Data* 56 (2011) 230–239.

[17] K. N. Semenov, N. A. Charykov, I. V. Murin, Y. V. Pukharenko, *J. Mol. Liq.* 201 (2015). 1–8.

In: Applied Aspects of Nanophysics …
Editors: Kirill Levine and Andrey Syrkov
ISBN: 978-1-53614-708-7

Chapter 49

REACTIVITY AND PROTECTIVE PROPERTIES OF SURFACE-MODIFIED DISPERSED ALUMINUM – PERSPECTIVE FILLER OF ORGANOPOLYMER COMPOSITIONS

***I. V. Pleskunov*[1], *V. R. Kabirov*[2], *Andrey Syrkov*[2,*] *and N. R. Prokopchuk*[3]**
[1]IMC Montan, London, UK
[2]Saint Petersburg Mining University, Saint Petersburg, Russia
[3]Belarusian State Technological University, Minsk, Republic of Belarus

ABSTRACT

The influence of alcamon (ALC) and triamon (TR) treatment on the oxidation rate of aluminum surface is reported. The synergetic effect which appears on the surface because of interaction of metal with nitrogen in the presence of quaternary ammonium compounds was studied using gravimetric and analytic methods. It was found that sequential modification of dispersed aluminum powder in vapors of different-sized molecules of ALC and TR leads to the significant (no less than 45%) increase of oxidation rate under high temperature (1173 K). The range of measurements of oxidation rate of differently treated samples is 0,012-0,019 g/m^2*min. According to XRD analysis, weak single peaks of aluminum oxide appears on diffractograms of Al/TR/ALC and initial samples. It is assumed, that amplification of oxidation of Al/TR/ALC is due to the following reasons. According to XPS analysis, the N1s bounding energy of nitrogen atom increases. This effect is associated with the steric availability of the nitrogen atom in triamon for its electronic interaction with metal and with the structural similarity of ALC and TR. As a result of amplification of heteroatomic interaction between metal and

* Corresponding Author's Email: syrkovandrey@mail.ru.

nitrogen (M←N), metal becomes more electronically saturated and demonstrates higher reduction properties in the air. The sample with triamon on the surface has the best anticorrosive properties out of all samples.

Keywords: quaternary ammonium compounds, tribology; metals treatment, adsorption modifying substrate, surface modification, metal reactivity, protective properties

INTRODUCTION

Earlier it was shown that as a result of a chemical sorption of different modifiers with donor and acceptor properties, tribochemical properties of different metals and systems on their basis can be significantly varied by solid-state reactions of metal reducing [1-3] or by modification of metal [4]. Except tribological properties, the water repellent properties and surface activity for participation in specific chemical reactions can be greatly varied as well. The following hydride-based surface modifiers were mainly used for surface treatment as reducing agents: NH_3, CH_4, SiH_4 [1, 2]. When Ni, Cu, Fe or Al surface was chosen as an adsorption-modifying substrate, best results for surface modification were obtained when substances based on quaternary ammonium compounds (QAC): triamon and alcamon, were used for modification [3, 5]. In addition, mentioned metals could be successfully used as filler of organopolymer compositions [3, 4].

Here it is reported the step-wise process of treatment of a dispersed Al in triamon and alcamon vapors, which results in a synergetic acceleration of metal oxidation in the process of its thermal treatment (1173 K) at a factor of not less than 45%.

EXPERIMENT

Al-powder PAP-2 (GOST 5494-71) with a unit of specific surface area 2.6 m^2/g was treated by vapors of TR and/or ALC at room temperature according to the developed methods [3, 5].

Compounds TR and ALC belong to liquid cationic surfactant with general structure $[R_1R_2R_3R_4N]X$, where R_i – organic group directly bonded to nitrogen atom in hydrophobic cation belonging to surfactant, X – inorganic anion [2]. The structural formula of TR, which possess lower molecular weight than ALC, is $[(HOC_2H_4)_3N^+CH_3][CH_3SO_4^-]$.

ALC possess methyl benzylsulfite polar group and alkyl radicals belonging to the structure of a cation, with a number of units n = 16-18).

X-ray fluorescent analysis (RFA) was carried out on "Bruker S4 Explorer" X-rays fluorescent spectrometer. Measurements were taken without filter at 10 kV voltage and 100 s exposure time.

Element dispersion X-rays analysis (EDX) was obtained by EDAS/TSL attachment of Nanolab electron scanning microscope. Surface sensitive mode with applied potential 6 kV was used for EDX measurements.

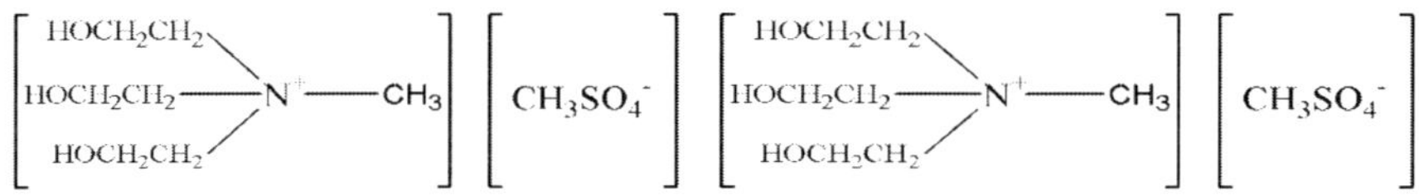

Figure 2. a) Triamon, b) Alcamon.

Results and Discussion

According to RFA analysis after the treatment in ALC vapors Al/ALC sample contained 0.13 at.% N and 0.12 at.% S; Al/TR sample after TR vapors treatment contained 0.21 at.% N and 0.22 at.% S. Step-wise Al powder treatment by TR and ALC vapors, sequentially, resulted in an increased N and S contents in Al/TR/ALC samples up to 0.55 and 0.43 at.%, respectively. According to X-ray fluorescent data and EDX-spectroscopy data, initial Al powder did not contain any noticeable amounts of nitrogen either sulfur, (Table).

Carbon contents in modified metallic samples did not exceed 2.7 at.%. Specific surface area of all the samples was approximately equal – 2.7 ± 0.1 m^2/g (BET). Obtained metallic samples with modified surface were simultaneously heated in muffle furnace (1173 K, 300 s) in air under (101 ± 1 kPa) pressure. Gravimetric measurements were taken to determine relative mass increase ($\Delta m/m$) during oxidation [2, 3], Figure 2.

Calculation of the rate of heterogeneous oxidation of samples V_{ox} was carried out according to a standard procedure [2]:

$$V_{ox} = \Delta m/(m \cdot S_{un} \cdot t)$$

where: S_{un} – specific surface area (m^2/g), t – time.

It was shown that surface modified metal oxidation rate depends on the kind of modification applied in this process and for samples Al/ALC bears the value 0.013 $g/(m^2 \cdot min)$, Al/TR – 0.012 $g/(m^2 \cdot min)$, Al/TR/ALC – 0.019 $g/(m^2 \cdot min)$, and for the initial Al-powder – 0.014 $g/(m^2 \cdot min)$. The highest Al/TR/ALC samples oxidation rate is confirmed by a precision instrumental analysis methods (EDX, XRD) on aluminum-based powders. Application of two-component (TR/ALC) layer on Al surface, significantly, (by no less than 1.45 times) exceeds an effect of each of this component taken separately.

According to data obtained by EDX, samples surface composition after heating under conditions specified above may be characterized as:

- Al/TR/ALC: Al – 82.1 at.%, O – 15.6 at.% (before oxidation: O – 7.1 at.%), C – 1.6 at.%, N – 0.39 at.%;
- Al/TR: Al – 88.2 at.%, O – 10.1 at.% (before oxidation O – 6.6 at.%), N – 0.18 at.%;
- Al/ALC: Al – 87.3 at.%, O – 10.7 at.% (before oxidation: O – 6.7 at.%), C – 1.5 at.%, N – 0.08 at.%.

Table 1. Content of elements adsorbed on Al-powder surface

Sample	Elements content			Molar ratio N/S
	N (EDX), at.%	S (EDX), at.%	S (RFA), mass %	
Al	-	-	-	-
Al/ALC	0,13	0,12	-	1,08
Al/TR	0,21	0,22	-	0,95
Al/TR/ALC	0,55	0,43	0,81	1,27
Al/(ALC+TR)	0,32	0,59	1,10	0,54

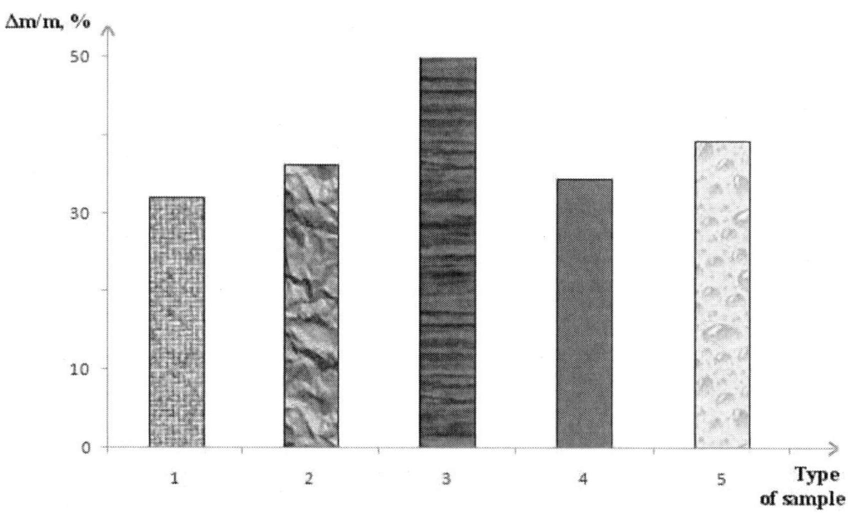

Figure 2. The influence of the composition of a treating layer on relative mass increase during oxidation (1773 K, 300 s) for Al-powder samples: 1-Al/TR, 2-Al/(ALC+TR), 3-Al/TR/ALC, 4-Al/ALC, 5-Al.

It is interesting that, according to XRD data, weak discrete peaks of aluminum oxide were observed only from solid Al/TR/ALC sample and from an initial Al powder.

Al/TR/ALC sample oxidation rate (V_{ox} = 0.019 g/(m^2·min) was noticeably higher than that of Al/(ALC+TR) sample obtained by treating by a mixture, as well as for an initial aluminum powder. Accuracy of the results is within 3% of calculated V_{ox} value.(?) Al powder chemical activity at oxidation and burning is considered to be compatible with the activity of other types of Al nanopowders [7]. It is assumed that the highest burning rate of Al/TR/ALC sample is related to a stabilization of bi-layered TR/ALC nanofilm and triamon underlayer at metal (M) surface due to formation of heteroatomic bonds M←N.

The possibility of mentioned metal-nitrogen interaction can be confirmed by increasing approximately by 2-3 eV N1s binding energy, according to XPS data [2, 3]. According to our understanding, this is due to steric availability of nitrogen atom in triamon for electronic interaction with metal and structural co-planarity of QACs in triamon and alcamon mixtures [3, 5, 9, 10]. As a result of nitrogen and metal surface interaction, the metal, obviously, becomes more electronically saturated system, and more actively behaves as reducing agent under conditions of oxidation in air.

CONCLUSION

Throw out selecting combination of QACs, synergistic effect can be achieved with a pronounced influence of surface chemical activity, such as oxidation rate. QACs treatment of metals has variety of practical applications in decreasing of friction in the environment of organopolymer lubricants [9], modifying surfaces for application as protective coatings, and corrosion protection.

ACKNOWLEDGMENTS

The work was carried out with the financial support of the Ministry of Education and Science of the Russian Federation (State contract № 14.577.21.0127 from 20 October 2014. Unique identifier of applied research RFMEFI57714X0127).

REFERENCES

[1] Syrkov, A. G., Pecularities of formation and structural and chemical features of metal products of hydride synthesis, *Rus. J. Gen. Chem,* 1994, Vol. 64, N.1, P. 43-50.

[2] Syrkov, A. G., Water repellent properties of the metal powders obtained by method of two-stage hydride synthesis, *Rus. J. Gen. Chem.*, 1995, Vol. 65, N. 11, P. 1920.

[3] Syrkov, A. G., *Nanotechnology and Nanomaterials. Surface-Structured Metals*, St. Petersburg: Polytechnical University Press (2012).

[4] *Modification of properties of elastomeric compositions*, N. R. Prokopchuk et al. Minsk: Belarusian State Technological University, 2012, P. P. 217.

[5] *Surface Phenomena and Surfactants*, Abramzon, A. A. and Schukin, E. D., Eds., Leningrad: Khimiya (1984).

[6] Gromov, A. A., A. P. Il'in, U. Foze-Bat, and U. Taipel About influence like the passivating covering, particle sizes and shelf-lifes on oxidation and nitriding of powders of aluminum, *Physics of Hot Explosion*, 2006, N. 2, P. 61-69.

[7] Gerasimov, Ya. I., *Kurs fizicheskoi Khimii* (*Textbook of Physical Chemistry*), Moscow: Khimiya Vol. 2 (1973).

[8] Makhova, L. V., A. G. Syrkov, I. V Stepanova, and A. V Fedotov About influence of binding energy of N1S of the adsorbed nanostructures on the greasing effect of surfactants on a interface metal-glass and metal-polymer *Condensed Matter and Interphases,* 2006, Vol. 5(4), P. 423-428.

[9] Nazarova, E. A., A. G. Syrkov, V. N. Brichkin Nonlinearity of dependence of integral friction index of tribosystem from hydrophilic properties of surface-modified metal fillers *Advanced Materials Research,* 2014, N.1040, P. 103-106.

[10] Syrkov, A. G. Synergetic change of tribochemical properties of copper in the presence of quaternary ammonium compounds at the surface. *Rus J. Gen. Chem,* 2015, Vol. 85(6), P. 1538-1539.

In: Applied Aspects of Nanophysics …
Editors: Kirill Levine and Andrey Syrkov
ISBN: 978-1-53614-708-7

Chapter 50

ADSORPTION OF WATER VAPORS ON DISPERSED COPPER CONTAINING DIFFERENT- SIZED MOLECULES OF AMMONIUM COMPOUNDS

V. R. Kabirov, Andrey Syrkov* and V. V. Taraban
Saint Petersburg Mining University, Saint Petersburg, Russia

ABSTRACT

The change of the water repellent properties on 24-hour time scale of dispersed copper, modified by quaternary ammonium compounds was analyzed in collation with tribochemical properties of the same samples as additives in industrial oil. It was established that the samples modified in mixed and consistent modes by both modifiers (alkamon (A) and triamon (T)) reach the adsorptive saturated state faster than others, due to the small number of hydrophilic centers on the surface of metals. Based on the exponential time dependences of water repellent properties, and measured integral index of friction of the samples, the trend of decreasing of integral index of friction with increasing of kinetic coefficient of adsorption in the series: Cu/T/A, Cu/T, Cu/A, was determined.

Keywords: chemisorption of quaternary ammonium compounds, copper, anti-frictional effect, surface of metal, tribochemistry, hydrophobicity, adsorptive saturation

* Corresponding Author's Email: syrkovandrey@mail.ru.

INTRODUCTION

Dispersed metals are widely used in various fields of science and technology, including electronics, tribotechnology and heterogeneous catalysis. Due to the high chemical activity of the particles, dispersed metals can be oxidized in air, primarily by water vapors. One of the promising methods of protecting, stabilizing and modification of surface of disperse metals is layering of different-sized molecules of quaternary ammonium compounds (QAC). The advantages of the method are the possibility of applying submonomolecular coatings, low vapor pressure of QAC, gas-phase modification at room temperature, and environmental friendliness of the method [1-3].

Previously, it was shown that the chemisorption of QAC-based cationic surfactants has a positive effect on water repellent and antifriction properties of metals, such as aluminum [2, 3]. However, there are more interesting objects for modification such as dispersed copper powders, not only from the point of view of surface chemistry, but also in case of variety of practical use. Also in previous researches for aluminum powder was established that high hydrophobicity of the samples leads to the high anti-frictional properties in industrial oil.

In this research, the rate of change of water repellent properties of dispersed copper with adsorbed QAC on the surface were studied in saturated water vapors (5-24 hours) in terms of their antifrictional properties as additives to the industrial oil. This topic is significant not only for regulation of different properties of surface of dispersed metals, but to identify the perspective of practical use of surface-modified copper powder.

MATERIALS AND METHODS

Copper powder PM-1 (GOST 4960-75) with specific surface area of 0.34 ± 0.02 m^2/g and with specific volume of monolayer of nitrogen of 0.08 ml/g was chosen as the initial sample of dispersed metal. More dispersed non-porous aluminum powder PAP-2 was also used to analyze the distribution spectra's of adsorption centers on the surface of metal samples. According to electron microscopy, the modification of the surface of metals does not lead to a significant change of the shape and size of the particles.

Metal powders were treated in vapors of alkamon (A), triamon (T) and hydrophobic silicone-organic liquid (HSL-94) at room temperature using various adsorption programs. The composition of the obtained samples with adsorbed ammonium or silicon-organic compounds was determined by EDX spectroscopy (analytical attachment EDAX/TSL, shooting mode - 6 kV). The binding energies of the electrons of chemical elements on the surface were determined by the X-ray photoelectron spectroscopy. Measurements of the X-ray photoelectron spectra were carried out on Escalab 220iXL (University of Leipzig).

The binding energy of the characteristic level was determined with accuracy of 0.1 eV. The water repellent properties of the samples were evaluated by parameter 1/a, which was calculated from the value of adsorption of water vapors (a). Adsorption of water vapors was measured gravimetrically (g/g) at relative vapor pressure of $p/p_0 = 0.98 \pm 0.02$ and temperature of 20 ± 2°C. The fact of adsorption of water vapors on metal was proved by the presence of the O1s peak with binding energy of 532.5 eV in the X-ray photoelectron spectra of samples [4]. The integral index of friction D, which is proportional to force and coefficient of friction, of the tribological pair (steel-steel) with the lubricant was determined via the acoustic emission method using the certified instrument ARP-11 at 20-300 kHz according to the procedure described in GOST 27655-88. The industrial oil I-20 was used as the lubricant; the modified copper powder was added into the oil, the copper content was equal in all the experiments. The mathematical description of the process was obtained using programs such as MathCad and MS Excel.

Alkamon and triamon are widespread surfactants based on quaternary ammonium compounds with the general formula of $[R_1R_2R_3R_4N^+]X^-$, where R_1, R_2, R_3, R_4 – organic radicals on a nitrogen atom; X^- - is a polar group, usually an inorganic anion. It should be noted that alkamon has larger organic radicals (C_{16}-C_{18}) than triamon (C_1, C_2). The composition of technical triamon correspond to the formula: tris- (β-oxyethyl) methyl ammonium methyl sulfate $[(HOC_2H_4)_3N^+CH_3][CH_3SO_4^-]$. The industrial hydrophobizator HSL-94, which was used in the research, in fact, was a reference model.

Results and Discussion

Modification of samples of dispersed copper was carried out according to the methods [2, 3, 5] and confirmed by EDX-spectroscopy. Due to the fact that the initial Cu powder did not contain any noticeable amount of nitrogen and sulfur, it can be assumed that their presence in modified powder is a result of chemisorption of QAC. Therefore the fact of chemisorption can be proved by the chemical composition of the samples. The content in initial powder is: copper at least 91.6 at. %, oxygen is 3.1 at. %, and carbon is 4.7 at. %. After treatment in alkamon vapors, the copper sample with the specific surface area of 0.35 m^2/g contained 0.2 at. % of nitrogen, 0.3 at. % of sulfur, 3.7 at. % of oxygen, and 5.8 at. % of carbon. The similarly prepared Cu/A (treated in triamon vapors) sample with specific surface area of 0.32 m^2/g contained 0.4 at. % of nitrogen, 0.4 at. % of sulfur, 3.7 at. % of oxygen, and 5.0 at. % of carbon. The Cu/T/A sample (prepared via sequential treatment in the triamon and alkamon vapors) with the specific surface area of 0.36 m^2/g contained 0.7 at. % of nitrogen, 0.8 at. % of sulfur, 3.1 at. % of oxygen, and 6.3 at. % carbon [5].

Based on the experiments of adsorption of water vapors by various samples, the parameter a (value of adsorption of water vapors) was converted into parameter 1/a, which characterizes the water repellent properties of the samples (Table 1).

The kinetic equation of mass transfer (1) for adsorption processes was used as the basis to describe the process of adsorption of water vapor on metal in time [6, 7], where n (t) - is the function of the amount of adsorbed water, k - is the kinetic coefficient of adsorption, t - is the time.

$$\frac{dn(t)}{dt} = -k \times n(t) \Rightarrow n(t) \sim e^{-kt}, where \;\; n(t) \sim a \tag{1}$$

After elementary mathematical transformations:

$$1/a \sim 1/n(t) \sim e^{kt} \tag{2}$$

The correlation dependencies, based on the experimental data were plotted in the MathCad (Figure 1). The part of the graph, which is parallel to the OX axis (time) corresponds to the stage of saturation of the sorbent. A mathematical description of the function 1/a = f(t) with sufficiently high coefficient of determination R^2 (Table 2) was obtained.

The shape of the curves of the dependences and equations are similar to relaxation curves for other types of hydrophobic adsorbents, which were studied in Institute of general physics in Moscow [7]. In our previous researches of dispersed aluminum, where the exposure time was significantly greater (more than 320 hours), the dependences of water repellent properties were approximated using a set of elementary functions (ln t, t^2 and Gaussian function) [8].

Table 1. Examples of conversion of value of adsorption a into parameter 1/a

Sample	Exposure time, hours							
	3		6		12		24	
	a, rel. un.*	1/a	a, rel. un.	1/a	a, rel. un.	1/a	a, rel. un.	1/a
Cu-initial	0,00437	229	0,00438	229	0,00655	153	0,00657	153
Cu/A+T	0,00197	509	0,00295	339	0,00393	255	0,00394	255
Cu/A	0,00598	167	0,00498	201	0,00497	201	0,00499	201
Cu/T	0,00401	250	0,00404	250	0,00501	200	0,00509	200
Cu/T/A	0,00399	251	0,00499	201	0,00598	167	0,00599	167
Cu/HSL	0,00293	341	0,00294	341	0,00391	256	0,00488	205

* Relative units.

Table 2. Approximated equations 1/a = f(t) and coefficient of determination R^2

Sample	Equation 1/a=f(t)	Coefficient of determination R^2
Cu-initial	$f(t) = 104 + 201e^{-0,126t}$	0,838
Cu/(A+T)	$f(t) = 241 + 543e^{-0,229t}$	0,982
Cu/T/A	$f(t) = 166 + 217e^{-0,310t}$	0,997
Cu/T	$f(t) = 191 + 95e^{-0,126t}$	0,838

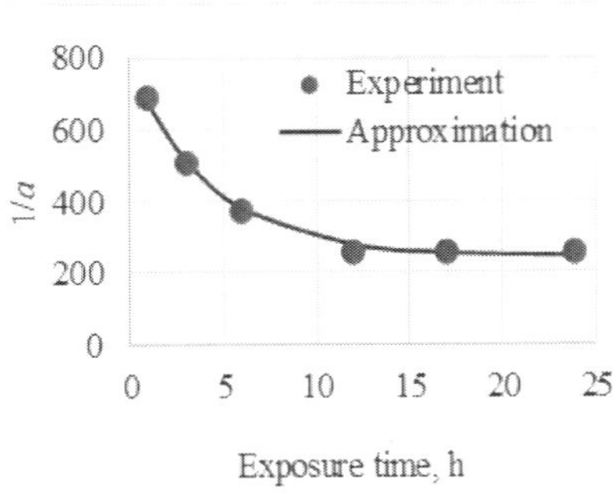

(a)

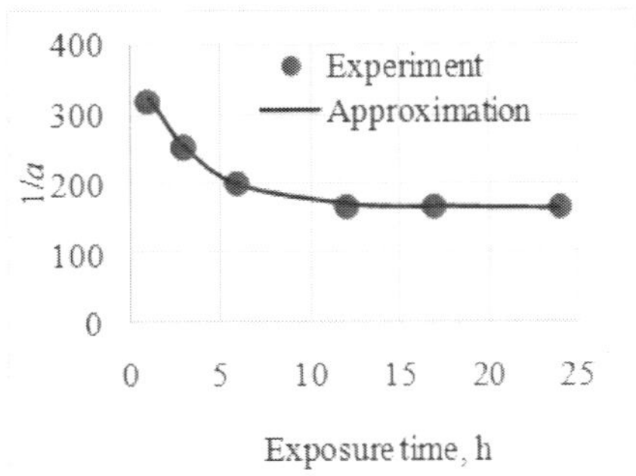

(b)

Figure 1. Dependency graphs 1/a = f(t) for Cu/A+T (a) и Cu/T/A (b) samples.

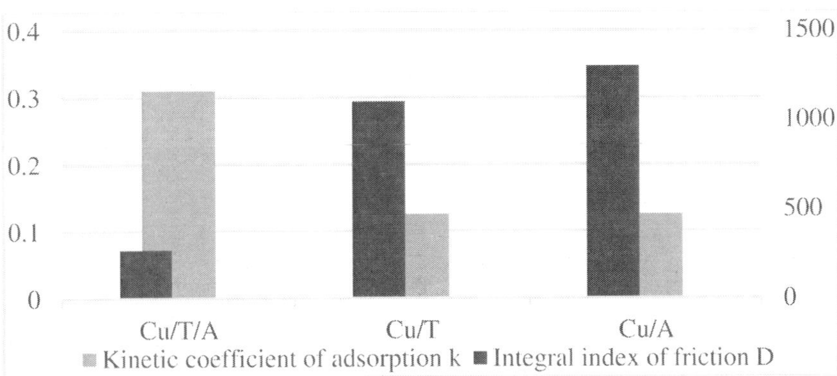

Figure 2. The relation of kinetic coefficient of adsorption and integral index of friction.

The sample Cu/(A+T) after 24 hours of interaction with saturated water vapor remains the most hydrophobic (the maximum values are 1/a (Table 1). According to Figure 1 and Table 2 samples treated with T and A in mixed and sequential modes (the maximum value of the coefficient k), reach the saturation state of the surface faster than the original copper or the sample treated with only one modifier (T).

Integral index of friction is a value proportional to force of friction and coefficient of friction. Measurement of the integral index of friction of the test system with the addition of the copper powders at the loading pressure of 40 MPa gave the following results: D1 = 1300 (Cu/A), D2 = 1100 (Cu/T), D3 = 270 (Cu/T/A), D = 1580 (initial copper), and D0 = 1690 ± 50 (no copper additive). According to the tribological results shown above and Figure 2, the kinetic coefficient of adsorption and integral index of friction are directly related to each other. Moreover, the highest value of k corresponds to the sample Cu/T/A with lowest integral index of friction D, which is the highly preferable. It is important to note that additives almost do not change their properties and composition in industrial oil during the time of experiment. The value k in Figure 2 was used in the form of absolute value.

CONCLUSION

For the first time the change of the water repellent properties of dispersed copper, modified using quaternary ammonium compounds (QAC) on 24-hour time scale on adsorption test in saturated water vapors was compared with integral index of friction.

Sufficiently accurate time exponential dependences of the water repellent properties of dispersed copper treated with QAC were derived and characterized. It was established that the samples, prepared by chemisorption of both modifiers in mixed and consistent regimes, reach the saturation state on the curves (1/a – time) faster than others, due to the small number of hydrophilic centers on the surface of metals. The sample Cu/T/A with the highest anti-frictional effect has the highest coefficient k (absolute value), which prove the stated above hypothesis from the tribology standpoint.

REFERENCES

[1] Syrkov A. G., Silivanov M. O., Kushchenko A. N. Tribochemical peculiarities of lubricant composition with surface-modified metal powder. *Journal of Physics: Conference Series*, 2016, Vol.729 p. 012026.

[2] Syrkov A. G., Kamalova T. G., Kabirov V. R., Kavun V. S. and Levine K. L. Influence of the triamon underlayers adsorpted on the interface on tribochemical characteristics of the metal-lubricant system, *Smart Nanocomposites*, 2015, N.6, p. 213-215.

[3] Syrkov A. G. Synergistic enhancement of the reactivity of aluminum in the presence of quaternary ammonium compounds on the surface *Russian Journal of General Chemistry*, 2013, N.83, p. 1392–1393.

[4] Roberts M, Mackie C. *Chemistry of the metal-gas interface* (M.: Mir), 1989, pp. 359.

[5] Syrkov A. G. Synergetic change of tribochemical properties of copper in the presence of Quaternary Ammonium Compounds at the surface. *Russian Journal of General Chemistry*, 2015, N.85, p. 1538–1539.

[6] Romankov P. G. *Methods of calculations of processes and apparatus of chemical technology* (SPb.: Khimizdat), 2009, pp. 544.

[7] Artemov V. G., Kurmasheva D. M., Kapralov P. O., Travkin V. D., Tikhonov V. I., Volkov A. A. Accelerated adsorption of water molecules with rapid contact with adsorbent Bulletin of the Russian Academy of Sciences. *Physical Series*, 2014, N.78, p. 245.

[8] Syrkov A. G., Kabirov V. R. Time dependence of the water repellent properties of the modified copper powder and the tradition of studying low-dimensional systems at the Mining University Materials of international forum-contest of young researches *Rational use of natural resources*, 2017, N.2, p. 236-237.

[9] Sychov M. M., Minakova T. S., Slizhov Yu. G., Shilova O. A., *Acid-base characteristics of the solid surface and properties control of materials and composites* (St. Petersburg: Chimizdat), 2016, pp. 276.

[10] Syrkov A. G., Silivanov M. O., Pleskunov I. V. Properties of the lubricants and its components containing surface-modified aluminum powder, *Smart Nanocomposites*, 2015, N.6, p. 171-180.

In: Applied Aspects of Nanophysics …
Editors: Kirill Levine and Andrey Syrkov

ISBN: 978-1-53614-708-7

Chapter 51

PECULIARITIES OF ZINC NANOPARTICLES FORMED ON GLASS AND QUARTZ SUBSTRATES BY ELECTRON SCANNING MICROSCOPY

Vladimir V. Tomaev[1,*] ***and Vladimir A. Polishchuk***[2]
[1]Saint Petersburg Mining University, Saint Petersburg, Russia
[2]Admiral Makarov State University of Martitime and Inland Shipping, Saint Petersburg, Russia

ABSTRACT

Zinc films obtained by magnetron sputtering on glass, and quartz substrates are investigated. The morphology of the surface, the thickness of the films, and the size of the zinc nanocrystals were studied by scanning electron microscopy. It is established that the morphology of the surface of zinc films essentially depends both on the technological application conditions and on the substrate material. Zn film formes very small islands on initial stages of their growth on quartz substrate, but formes more significant islands on glass substrates. The geometric and thermodynamic parameters of the investigated films were estimated. Resistance of zinc films was investigated by impedance spectroscopy. The conclusion is drawn that the conductivity of the films is of a percolation nature. With increasing film thickness is observation the twinning of Zn crystals.

Keywords: zinc films, magnetron sputtering, glass and quartz substrates

---

* Corresponding Author's Email: tvaza@mail.ru.

INTRODUCTION

This paper reports some results relating to the thin Zn films formed by magnetron sputtering on glass and quartz substrates. The problems of obtaining thin layers of metals, both from a practical and theoretical point of view, have long been studied [1-5]. Mechanisms of growth of metal films were considered in detail in [6, 7, 8, 9]. The structure and morphology of the films determine their electrophysical, optical and other properties.

During the spraying of metals, there are three types of crystal structures of the films: a hexagonal close-packed (HCP), face-centered cubic (FCC) and body-centered cubic (BCC) [6].

The basic process, which determines a film structure is heterogeneous nucleation. Volmer introduced the concept of "critical phase (embryo)" [6]. This is a particle of a new phase, for which the maximum work of formation is required. Thus, depending on the deposition rate, the substrate material and the type of metal to be deposited, there are three types of film growth [9]: island growth mode Volmer-Weber [6]; layer growth mode or the Frank-van der Merwe [8]; layer-by-layer growth, or mixed growth in the Stransky-Krastanov regime [9]. That is, what type of growth is currently advantage determined by the interaction of the film with the substrate atoms and atoms together. An island growth mechanism is realized when the condition [10]:

$$\sigma_s < \sigma_d + \sigma_{s-d} - const \times k_B T ln(\xi + 1) \quad (1)$$

where σ_s - free energy per unit surface of the substrate, σ_d - free energy per unit surface film, $\sigma_{(s-d)}$ - the free energy per unit surface substrate-film section.

EXPERIMENTAL

Zn films were deposited on glass, quartz and silicon substrates measuring 20×20 mm^2 by magnetron sputtering at various currents and constant pressure Ar ~1×10^{-2} Torr. The source was a target of Zn with a diameter of 57 mm. The distance between the target and the substrates, which are fixed on a rotating holder holding at a constant speed, is ~5 cm. The deposition rate was measured. The film thickness was determined by an integrated meter, the value of which was taken as the "effective film thickness." In this case the effective thickness is equal to h_{ef} film:

$h_{ef} = m/d\times S$,

where m - mass of deposited metal, d - the density of the metal material, S - area occupied by the film on the measuring surface.

As can be seen from the SEM results in Figure 1 a, b and the histograms of the size distribution of islands of the film structure on different substrates differ greatly.

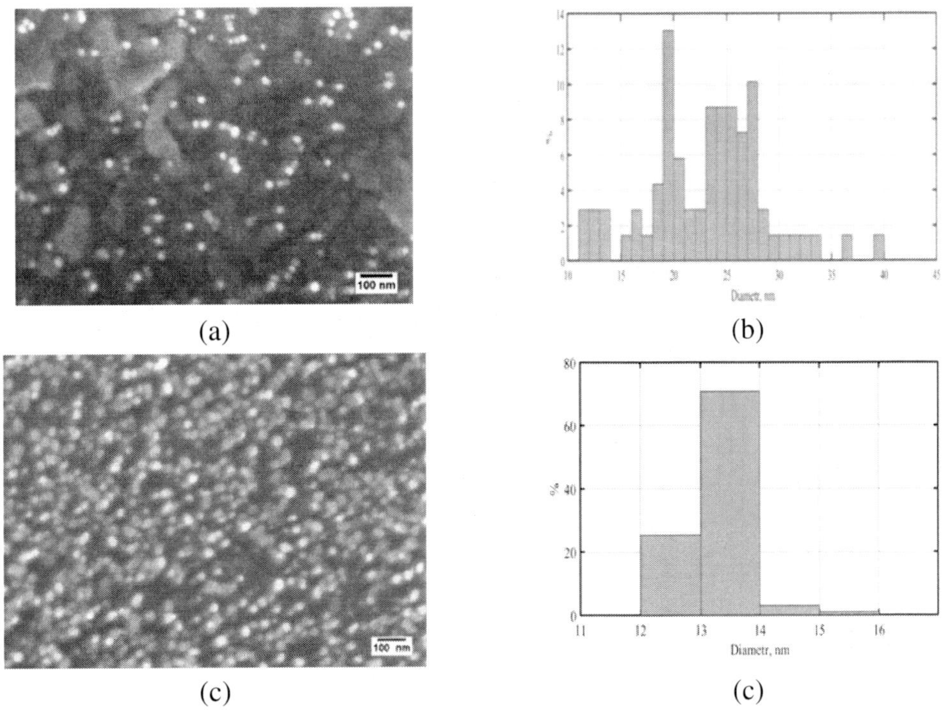

Figure 1. SEM view of the film structure on glass substrate (a) and quartz substrate (b).

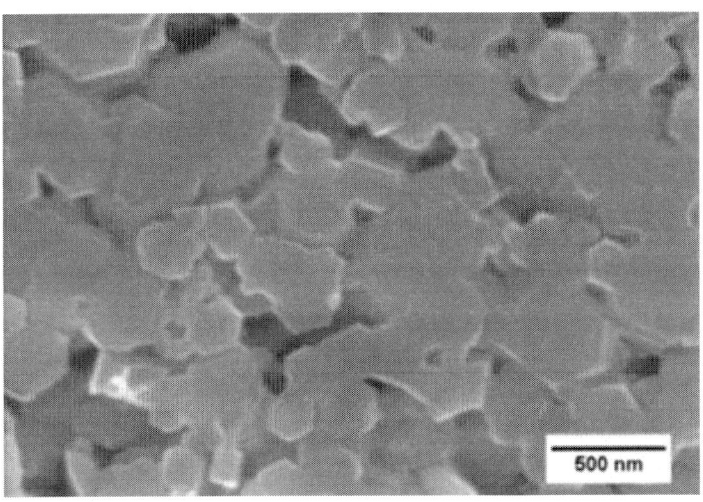

Figure 2. SEM view of the back side film.

Figure 3. SEM view of the "front" side film.

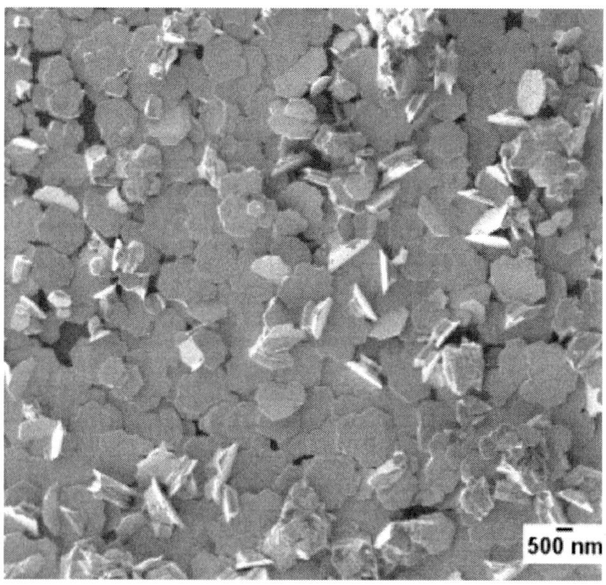

Figure 4. SEM view of the Zn film with an "effective" thickness 200 nm.

The average grain size on glass is 23.1 ±5.6 nm, on quartz 12.9 ± 0.6 nm. New atoms come from the vapor of the material during the sputtering of the target. Assuming that the nanocluster is a monolayer of Zn atoms, it consists of 300-1000 atoms [11]. With a further increase in the thickness of the Zn film, the size of the islands grows on a glass substrate. The morphology of the surface does not change significantly up to a thickness of 35 ± 40 nm. In Figure 2 can be seen of the back side of the film in which all hexagonal zinc crystals have an orientation in which the reference plane [0001] is parallel to the plane of the substrate.

As can be seen, each critical nucleus of a new phase that has reached a critical size begins to grow in accordance with the growth rules for the hexagonal crystal structure of

Zn in the form of a thin plate parallel to the plane of the substrate perpendicular to the axis from the hexagonal lattice. In Figure 3 shows the "front" side of the Zn film. It can be seen that individual zinc single crystals have a hexagonal shape and are oriented with respect to the surface of the substrate randomly. This orientation of single crystals of zinc hexagonal form is caused by a significant film thickness (of the order of 1000 nm). In Figure 3 can be seen that on all six faces of zinc single crystals, which are perpendicular to the base of the regular hexahedron, one can observe the "hatching" characteristic of polysynthetic twins. Figure 4 shows a Zn film with an "effective" thickness of about 200 nm. As can be seen, the main part of hexagonal zinc crystals has an orientation at which the [0001] plane is parallel to the plane of the substrate.

Figure 5 shows the dependence of the conductivity of Zn films on their thickness on quartz substrates.

The dependence of the film conductivity on the thickness is linear.

As the film thickness increases, more zinc nanoparticles begin to touch each other and an increasing number of zinc grains begin to participate in the conductivity of the film, which ultimately leads to a decrease in film resistance.

For thick zinc films (h > 100 nm), the conductivity of the film will tend to conduct the bulk polycrystalline zinc.

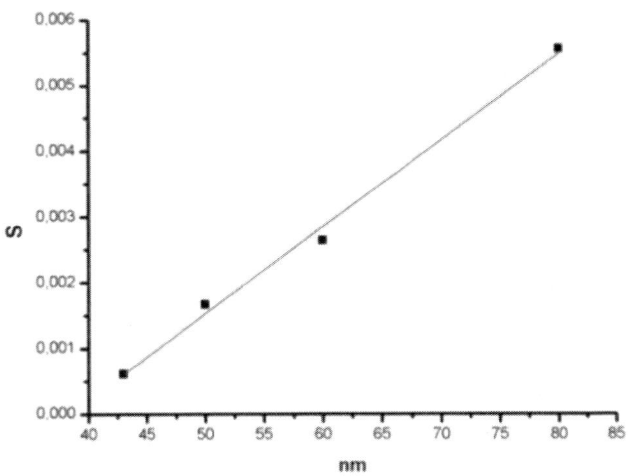

Figure 5. Dependence of the conductivity of Zn films on their thickness on quartz substrates.

DISCUSSION

The main process that determines the structure of the film is heterogeneous nucleation. Volmer introduced the notion of the "critical nucleus" of a new phase [6]. This is a particle of a new phase, for which the energy of growth is maximal. It is clear that the formation of the "critical nucleus" of a new phase will play a major role in the structure of the substrate. If the interaction between the adsorbed atoms is weaker than the interaction with the substrate atoms, should form large clusters for the formation of monatomic layers. If the interaction between the adsorbed atoms prevails, the formed clusters of atoms adsorbed quickly transferred critical nucleus in the new phase. Noteworthy is a sharp contrast to the distribution of Zn islands sizes on different substrates. The scatter in the sizes Zn islets, which is observed on a substrate made of glass (Figure 1a) can be explained by the surface irregularities, the surface of all possible structural defects. Under such conditions, the nucleation centers may be all kinds of impurity atoms and the surface contamination. On a substrate of quartz observed minimal scatter in the sizes Zn islets. This can be explained by the more "smooth" surface of quartz and the identical rate of growth of the embryos. In Figure 1b outlines of large size quartz crystallites. The islands of Zn are mainly located along the boundaries of sections of crystallites. Places of primary nucleation can include cleavage steps, dislocation outlets and etc. Nucleation clusters preceded by the formation of the adsorbed atoms and their growth to a critical size. The influence of the substrate on the formation of embryos will be manifested [12]:

- if the interaction between adsorbed atoms is weaker than their connection with the atoms of the substrate;
- if the interaction between adsorbed atoms is stronger than their connection with the atoms of the substrate.

In the first case, large clusters should arise, possibly the formation of monoatomic layers, in the second case small clusters of adsorbed atoms rapidly transfer to the nuclei of the new phase. We calculate the size of the critical nucleus. The expression for the critical nucleus size is:

$$R_0^* \approx \frac{\Omega \gamma_2}{KTln\left(\frac{P}{P_e}\right)}$$

where Ω - is the volume occupied by one atom in the cluster on the substrate, γ_2 - is the specific free energy for a given metal, (P/P_e) - is the critical supersaturation (of the order of 50 - 100). For Zn adatom volume of Ω=11.994 Å^3, γ_2 = 760 mJ/m^2, at a substrate temperature of 400 K° and satiety parameter (P/P_e) = 50 the radius of the critical nucleus is given by: R_0 = 4.1903 Å≃ 0.4 nm.

The obtained value of the critical nucleus radius R_0 smaller than for the other metals - for Ag [12] R_0 = 10.6 Å, - for Ni R_0 = 7.4 Å.

CONCLUSION

1) It is shown that on substrates of glass and silicon during magnetron sputtering, formed island films with an effective thickness of 10÷12 nm and an average grain size of 23.1 ±5.6 nm on glass.
2) On the substrate of quartz, island films with an average size of 12.9 ± 0.6 nm islands are formed.
3) Formed oriented hexagonal Zn single crystals with the [001] direction parallel to the plane of the substrate.

REFERENCES

[1] Levinstein H. 1949. *J. of Appl. Phys.* V 20 April 306.

[2] Konozenko I. D. 1954. *Physics-Uspekhi* V 52 N4 561.

[3] Yoo M. H. 1981. *Metallurgical Transactions A,* V 12A N3409.

[4] Kukuschkin S. A., Osipov A. V. 1998 *Physics-Uspekhi* V 168 N10 1083.

[5] Ohring M. 2002. *Material science of thin films Deposition and Structure*. Academic Press. USA.

[6] Volmer M., Weber A. 1926. *Z. Physik. Chem.* 119277.

[7] Stranskiy I. N, Kaischev I. K. 1939. *Physics-Uspekhi* V 21 N4 408.

[8] Frank F. C. and van der Merwe J. H. 1949 One-dimensional dislocations. I. Static theory, *Proc. Roy. Soc. London, Ser. A* 198205.

[9] StranskiI. N., Krastanow L. 1938. *Abhandlungen der Mathematisch-Naturwissenschaftlichen Klasse IIb* [*Treatises of the Mathematical-Scientific Class IIb*]. Akademie der Wissenschaften Wien. 146797.

[10] Venables J. A., Spiller G. D. T., Hanbücken M. 1984. *Rep. Prog. Phys.* 47399.

[11] Ievlev V. M., Shvedov E. V. 2006. *Physics of the Solid State V* 48 N1133.

[12] Tochitsky E. I. 1976. *Crystallization and heat treatment of thin films.* Minsk. BY.

In: Applied Aspects of Nanophysics …
Editors: Kirill Levine and Andrey Syrkov
ISBN: 978-1-53614-708-7

Chapter 52

THE AMBIENT PRESSURE APPROACH OF MODIFYING ZINC FILMS BY CHEMICAL AND PHYSICAL METHODS

***Vladimir V. Tomaev*[1,3], *Vladimir A. Polishchuk*[2], *Nikolai S. Pshchelko*[3], *Kirill Levine*[1] *and Sergey G. Zverev*[4]**

[1]Saint Petersburg Mining University, Saint Petersburg, Russia
[2]Saint Petersburg University of Information Technologies, Mechanics and Optics, Saint Petersburg, Russia
[3]Military Academy of Communications, Saint Petersburg, Russia
[4]St. Peterburg Polytechnical University, Saint Petersburg, Russia

ABSTRACT

The method of thermal treatment of zinc films in non-pressurized volume was developed and evaluated. Zinc films on glass substrates were subjected to thermal treatment in the presence of selenium vapors in air atmosphere, and combined with thermal-electric field treatment. Morphology was assessed by SEM. Structural changes in initial and modified Zn films were discussed. Size and statistical parameters of zinc nanocrystals were determined. It was possible to control zinc films' morphology and composition in a wide range via the suggested method.

Keywords: zinc films, surface morphology, nanocrystals of zinc and zinc selenide, selenium vapors, new phase nucleation

INTRODUCTION

Zinc selenide (ZnSe) is known for its photo-resistive, photo- and electro-luminescent properties. Due to its high transparency in optical and infrared regions, it is used for manufacturing entry lenses and windows in optoelectronic devices [1].

The problem of obtaining, studying and usage of nanocrystalline films based on zinc halogenates is on the edge of study [2-4].

In [5] it is realized the idea of optimization of photosensitive lead selenide (PbSe), by heating precursor films in the presence of selenium powder in air atmosphere in a non-hermetically closed vessel. As a result of this procedure it was possible to obtain PbSe films with infrared sensitive capability at ambient temperatures (without cooling).

Polycrystalline ZnSe can be obtained by interaction of zinc with selenium vapors [5]. Monocrystals of cubic singonia were grown from vapor phase and melt, while monocrystals of hexagonal singonia – only from the vapor phase. ZnSe films were obtained by thermal evaporation of the compound and condensation on a substrate heated to 150-250°C.

METHOD

Method in [5] describes chemical precipitation of PbSe from multicomponent mixtures under next following conditions:

- films are placed at non-hermetically close vessel, where certain ratio between surrounding volume and surface area of heating films is maintained, (usually ratio of total surface area to treating is 20 - 30).
- heating is carried out in the presence of metallic selenium.

Heating of Se in powder form allows saturating vessel with vapors of high partial pressure. Diffusion at temperatures of treatment creates conditions of controlled Se penetrating to lead. Varying experimental conditions allows varying Se concentration, decreases percentage of overoxidated sites and simultaneously decrease number of Se vacancies in PbSe crystalline lattice.

EXPERIMENTAL SECTION

Depositing Selenium on Zinc

In this paper same method as described above was applied to treat zinc films by Se vapors.

Se surface-modified Zn films were obtained as a result of chemical interaction of Zn surface with Se vapors.

Isobaric conditions from one side, isolating from ambient atmosphere from the other, effectively decrease amount of oxygen during the reaction with simultaneous increasing of Se partial pressure. Oxygen becomes effectively consumed at initial stages of treatment and does not interfere Se deposition further. Oxygen partial pressure decreases 2.0 – 2.2 times during the reaction, while Se vapors pressure competitively increases. Partial remove of oxygen from non-hermatically sealed volume and increasing concentration of Se vapors occurs due to higher Se vapors pressure comparatively to oxygen pressure at given temperature. Therefore, in non-hermetically closed vessel, same effect as for hermetically closed is achieved, while complexity of experimental set-up is significantly decreased.

Considering high ionic contribution to chemical bonding, low thermal conductivity, high non-stoichiometry, polymorphic trend, high melting temperature, halogenates synthesis by classical methods meets certain experimental difficulties [4].

Therefore, alternative halogenate synthesis methods, especially with surface modified on nano-level, is of interest both from fundamental, and applied side.

In this paper it is suggested method of thermal treatment in selenium vapors. Applied to thin film coatings, this method allows creating nanometer-size selenium structures integrated to coating and chemically bonded to it.

Thermo-Electric Treatment of Zinc

Zinc films of ~ 1000 nm thickness were obtained by vacuum thermal evaporation method on glass substrates (Figure 1).

For the majority of crystals, hexagonal structure is typical, which is indicative for rather high purity of grown Zn crystals and conditions of either homogeneous, either heterogeneous conditions of crystalline nucleation.

Cross-sectional Zn crystalline dimensions range from 200 to 1000 nm for films of nearly 1000 nm thickness.

Zinc crystals were heated to 250°C for 2 hours in the presence of selenium powder in non-hermetically closed vessel by method analogous to [6].

Figure 1. Surface morphology of zinc films obtained by thermal evaporation onto glass substrate.

As a result of treatment in Se vapors, Zn crystalline substrate was coated by nanometer (~ 20-50 nm) size particles (Figure 2).

Comparison of morphologies of untreated (Figure 1) and treated (Figure 2) in Se vapors Zn films shows that at treated film surface morphological features of a smaller size (20 – 50 nm) are formed, which were not present at initial film.

Also, at the edges of hexahedrons on perimeter protrusions can be noticed. This phenomenon can be explained by interacting between Zn and Se vapors.

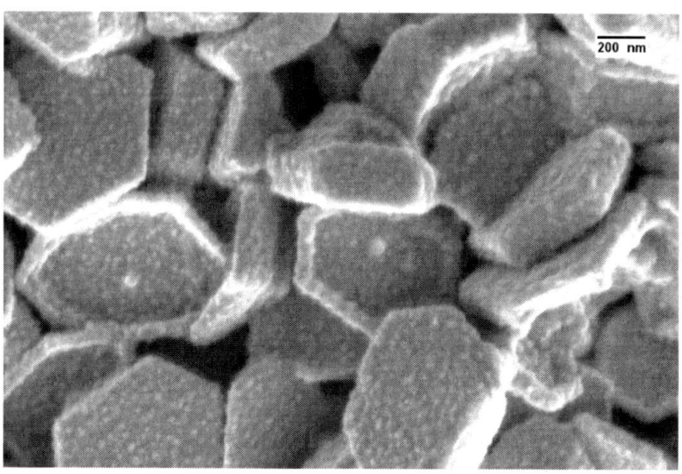

Figure 2. Zinc crystals after treatment with selenide vapors.

Another advantageous way of surface modification is developing by us method, where simultaneous temperature and electrical gradient in air atmosphere results in promoting electro-adhesion effects and creating films with significant differences in structure from ones synthesized without electrical field.

Zn film on glass was modified by constant electrical potential U = 300 V, exposition time was 10 min and temperature 250°C. Positive potential was applied to Zn. Precursor sample was placed in a holder that at the same time was upper electrode. Simultaneous effect of thermal treatment and electric field was studied as in [7, 8]. Control sample was placed near treated sample, without applied voltage.

The influence of temperature and electric field on adhesion and phase composition of metallic zinc films was studied by experimental setup shown in Figure 3. Experiments were performed in dry air atmosphere. Films for deposition were deposited on glass substrates.

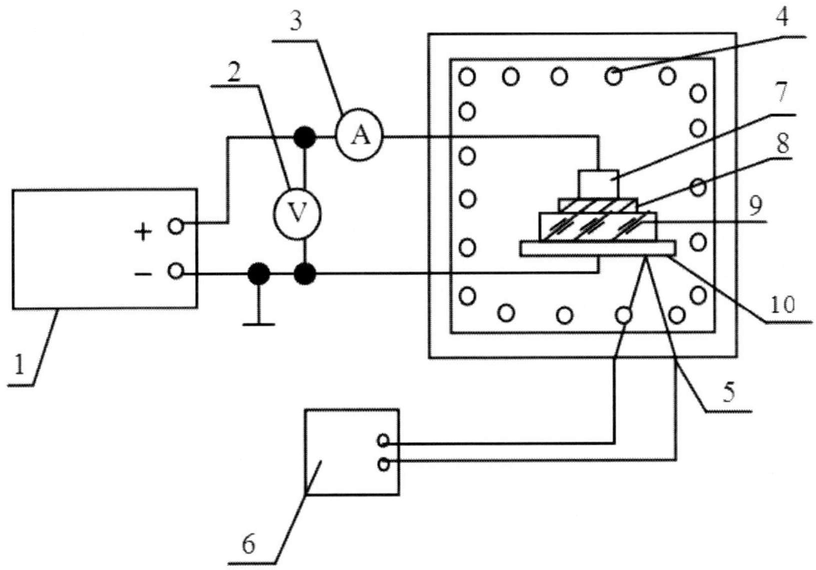

Figure 3. Experimental setup for electro-adhesion experiments. 1 – powere supply. 2 – voltmeter, 3 – current recorder. 4. Oven, 5 – thermo-couple, 6 – high sensitive potential meter, 7, 10 electrodes, 8 – substrate for deposition, 9 – dielectric plate.

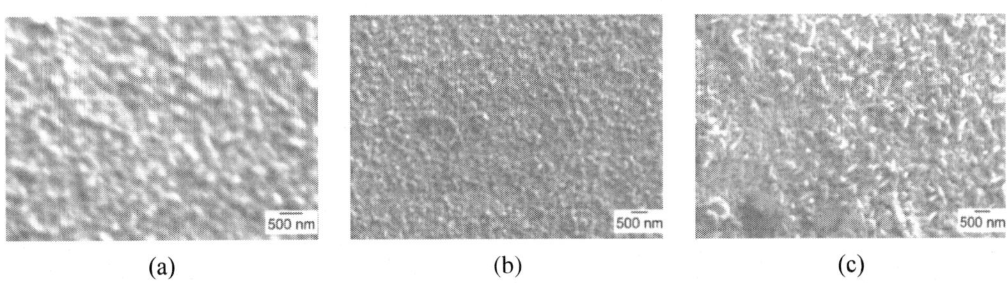

Figure 4. Treated and untreated films SEM images. (a) precursor film, (b) film affected by thermal treatment in air, (c) films thermally treated with applied cross-sectional electric field. Exposure time 10 min, temperature 250 C°.

Treating sample with zinc film was hold in a device serving upper electrode at the same time. Thermal treatment with simultaneous action of electric field was studied. Control sample (outside of the electric field) was placed nearby. Modification of film at glass substrate was handled at U=300 V, treatment time was t = 10 min and T = 250°C.

Obtained results show increased surface resistance ($R_1 \approx 15$ Ohm/sq) comparing to untreated sample ($R_2 \approx 4$ Ohm/sq). This is probably due to Zn oxidation by oxygen ions electro-migrating from glass to Zn. Transition from not-transparency to semi-transparency confirms this assumption. As known, ZnO are optically semi-transparent in visible band.

As a result of complex (temperature-electric) treatment, partial Zn oxidation to ZnO occurs. Large-gap (3.37 eV) semiconductor is a high-ohmic semiconductor with resistivity ~ 10^{-3} Ohm·cm) at room temperature. Partial Zn oxidation at applied experimental conditions results therefor in a significant decrease in conductivity, which is for pure Zn ~ $5{,}9 \cdot 10^{-6}$ Ohm·cm).

Surface morphology was studied by scanning electron microscopy, while surface composition was evaluated by elemental analysis. Those measurements were carried out on Zeiss Merlin) electronic microscope. Figure 4 shows microphotographs of films surfaces obtained at different stages of treatment.

Figure 4 shows that surface consists of separate, adjacent to each other crystals, which size ranges from 50 to 200 nm. From Figure 4 b) it is clear that temperature does not strongly affect film structure, resulting only with its partial oxidation. Fragment of Zn film surface after 10 min temperature treatment in electric field. Figure 4 c) is featured by large amount weakly oriented towards surface one-dimensional crystals. While cross-sectional size of those crystals is 50…100 nm, their length is 500… 1000 nm. The majority of filament-shape crystals are oriented towards surface randomly. Therefore, electric field significantly changes film structure: non-uniform electric field favors obtaining ZnO films with developed surface. Those ZnO films are composed from one-dimensional crystals of nanometer size.

CONCLUSION

Preliminary studies have shown that by varying technological regimes of thermal treatment under constant electric voltage, this is possible to alter concentration and linear dimensions of linear zinc oxide crystals.

Metal was converted to a semiconductor by treating in the presence of oxygen and an electric field.

Technological findings obtained in this study are believed to be useful for micro fabrication appliances, where expensive vacuum technologies need to be replaced by more cost-effective surface treatment methods.

REFERENCES

[1] Dubey, JP; Tiwari, RK; Upadhyaya, KS; Pandey, PK. Crystal dynamics of zinc chalcogenides I: an application to ZnS. *Turk J Phys*, (2015), 39, 242 – 253, http://journals.tubitak.gov.tr/physics/.

[2] Dubey, JP; Tiwari, RK; Upadhyaya, KS; Pandey, PK. Crystal dynamics of zinc chalcogenides II: An application to ZnSe. *IOSR Journal of Applied Physics (IOSR-JAP)*, e-ISSN, 2278-4861, Volume 7, Issue 5 Ver. I (Sep. - Oct. 2015), PP 67-75 www.iosrjournals.

[3] Panchal, CJ; Opanasyuk, AS; Kosyak, VV; Desai, MS; Protsenko, IYu. Structural and substractural properties of zinc and cadmium chalcogenides thin films. *J. Nano-Electron. Phys*., 3, (2011), No 1, p. 274-301.

[4] Triboulet, R. Growth of Zinc Chalcogenides/In book: *Crystal Growth Technology*, 2003, pp. 497-523, DOI: 10.1016/B978-081551453-4.50017-8, http://www.sciencedirect.com/science/article/pii/B97808 1551 4534500178#.

[5] Mironov, AP; Markov, VF; Maskaeva, LN; Diakov, VF; Mukhamediarov, RD; Mukhamedzianov, HN; Smirnova, ZI. *Method of sensibilization of chemically precipitated films of lead selenide for IR radiation*. Patent RU 2357321.

[6] *Database JCPDS No. 03-065-5973.*

[7] Pshchelko, NS; Sevryugina, MP. Modeling of physical and chemical processes of anodic bonding technology. *Advanced Materials Research*, Vol. 1040, 2014, p. 513-518.

ABOUT THE EDITORS

Dr. Kirill Levine

Docent
Department of General and Technical Physics
Saint Petersburg Mining University,
Saint Petersburg, Russia
Email: levinkl@hotmail.com; levin_kl@pers.spmi.ru

Dr. Kirill Levine graduated from the radiophysics faculty of Leningrad's Polytechnic University. He obtained his PhD at the University of Cincinnati (UC) (Ohio, USA) at the Department of Materials Science and Chemical Engineering. Dr. Levine worked as a postdoc at North Dakota State University (NDSU) (North Dakota, USA). He currently holds a position of Docent at St. Petersburg Mining University since 2010. Area of scientific studies: corrosion protection, nanocomposites formulation.

Dr. Andrey Syrkov

Professor
Department of General and Technical Physics
Saint Petersburg Mining University, Saint Petersburg, Russia
Email: syrkovandrey@mail.ru; syrkov_ag@pers.spmi.ru

Since 2001, Dr. Andrey Syrkov has been docent of the Department of General and Technical Physics of Mining University. Since 2003 he has held the position of a Professor, which was certified at 2006. Since 2012 is a director of scientific and research center "Nanotechnologies" at Mining University.

INDEX

N

O

P

Q

R

S

T

U

V

W

Z